Functional Materials

Edited by Dipti Sahu

Published in London, United Kingdom

IntechOpen

Supporting open minds since 2005

Functional Materials
http://dx.doi.org/10.5772/intechopen.76823
Edited by Dipti Sahu

Contributors
Denisse Ortiz-Acosta, Tanya Moore, Sabah Abdelbasir, Ahmed Esmail Shalan, Jiyong Fu, Wei Wang, Minghua Zhang, Nor Hapishah Abdullah, Muhammad Syazwan Mustaffa, Mohd Nizar Hamidon, Farah Nabilah Shafiee, Rabaah Syahidah Azis, Jeetendra Kumar Banshiwal, Durgesh Nath Tripathi

First published in London, United Kingdom, 2019 by IntechOpen
IntechOpen is the global imprint of INTECHOPEN LIMITED, registered in England and Wales, registration number: 11086078, The Shard, 25th floor, 32 London Bridge Street
London, SE19SG - United Kingdom
Printed in Croatia

British Library Cataloguing-in-Publication Data
A catalogue record for this book is available from the British Library

Additional hard and PDF copies can be obtained from orders@intechopen.com

Functional Materials
Edited by Dipti Sahu
p. cm.
Print ISBN 978-1-78984-057-5
Online ISBN 978-1-78984-058-2
eBook (PDF) ISBN 978-1-83880-045-1

For EU product safety concerns:
IN TECH d.o.o., Prolaz Marije Krucifikse Kozulić 3, 51000 Rijeka, Croatia, info@intechopen.com or visit our website at intechopen.com.

Meet the editor

Prof. Dipti Sahu is currently an associate professor of physics in the Department of Natural and Applied Sciences at the Namibia University of Science and Technology. He received his doctoral degree in Physics from the Institute of Materials Science, Utkal University, India. After his PhD, he worked as a postdoctoral researcher, visiting scientist, and faculty in several institutions such as the National Taiwan University, National Cheng Kung University (Taiwan), and the University of Witwatersrand (South Africa). His research focus is on multifunctional materials, including nanomaterials, ceramics, composites, spintronics, magnetic materials, and the application of functional materials in devices. He has published more than 85 international peer-reviewed journal articles, two book chapters, and more than 100 research articles in conference proceedings and meetings.

Contents

Preface

Functional materials are highly promising advanced materials that exhibit many interesting phenomena directly related to device applications. Development of functional materials and their effective applications in the frontier fields of cross-multidisciplinary research programs ranging from earth science to physical, chemical, biological, space, and information science are being increased many fold. Rapid fusions of the above-mentioned disciplines require innocuous and high value-added advanced materials for cross-disciplinary science and technologies spanning from single crystal bulk design to thin and ultra-thin film structures of single or multi-elemental material compositions in the form of metals, alloys, ceramics, and composites. These materials are classified based on functions that are significantly different to previously described materials based on their origin. New functional materials and composites can be developed using new fabrication techniques and characterized using advanced characterization methods. The relationships between processing, structure, properties, and performance will be explored and effective functional applications for devices will be mentioned. This book provides an overview of different types of functional materials, including synthesis, characterization and application, and up-to-date treatment of functional materials, which are needed for structural, magnetic, polymeric, electromagnetic, etc. applications. It is suitable for undergraduates, graduates, and professionals, including engineers, scientists, researchers, technicians, and technology managers.

Dipti Ranjan Sahu
Namibia University of Science and Technology,
Windhoek, Namibia

Section 1

Functional Polymeric Materials

Chapter 1

Functional 3D Printed Polymeric Materials

Denisse Ortiz-Acosta and Tanya Moore

Abstract

Additive manufacturing (AM) is an emerging 3D printing technology that enables the design and rapid manufacturing of materials with complex microstructures. Advances in 3D printing have allowed manufacturing companies to expand from design and 3D printing of prototypes to the rapid manufacturing of end products. Additive manufacturing enables the manufacturing of components in a layer-by-layer fashion, opposite to common manufacturing methods that rely on machining, molding and subtractive methods to obtain the final product. AM employs a computer-aided design software that allows for the design of virtual objects and the control of the nozzle and/or stage of the 3D printer. Due to their versatility and wide range of mechanical and chemical properties, polymers are the most utilized materials for AM. Polymers used for AM covers thermoplastics, thermosets, elastomers, polymers with incorporated fillers, biopolymers, and polymers blended with biological materials. The architectural design and choice of polymers can lead to materials with enhanced functionalities, mechanical properties, porosity, and stability. This chapter focuses on the development of polymer-based 3D printing materials with multifunctionalities used specifically for the production of biomedical devices, electronic devices, and aerospace-relevant products.

Keywords: 3D printing, additive manufacturing, polymers, biomedical devices, aerospace, electronics

1. Introduction

3D printing is an additive manufacturing (AM) process that enables the manufacturing of components with complex geometries in a layer-by-layer fashion. 3D printing became popular after the first machine was introduced to the market in 1986 by Hull [1]. Charles Hull created the first stereolithography (SLA) manufacturing method which he used for the rapid design and manufacturing of small prototype plastic parts. Stereolithography uses light to activate polymers within a resin (photopolymerization) to create 3D, complex shapes [2, 3]. This SLA system was commercialized in 1987 by the company 3D Systems. Since this breakthrough invention, there has been great effort in producing machines that can process a variety of plastics. Some of the machines currently in the market are fused deposition modeling (FDM) [4, 5] and direct ink write (DIW) for extrusion-based processes [6, 7]. Powder bed fusion (PBF) and laser sintering (SLS) are used for processes requiring a laser to cure or fuse polymeric materials [8]. Inkjet printers also use light to photopolymerize ink drops into complex shapes [9]. Extensive reviews on these processing and 3D printing technologies have been published elsewhere [4, 5, 10–14]. This chapter focuses on applications that use AM for the 3D printing of polymeric materials.

2. 3D printing general process

Since the 1980s, 3D printing has become very popular as a result of the rapid manufacturing of components with architectures designed to meet specific applications. AM allows for the manufacturing of a variety of shapes in a layer-by-layer fashion, often without the need of post-processing such as machining. As a general scheme, AM starts with the design of a virtual object using CAD (computer-aided design) software that generates a STL (stereolithography, named after Charles Hull's SLA process) file format [15]. A slicer program interprets the STL file and converts it into g-code (e.g. Slic3r, 3DPrinterOS, MakerBot Print, and others). The computer controls the stage and dispenser of the 3D printer allowing prototypes to be manufactured. Rapid prototyping allows one to refine product ideas while saving significant time and money because it allows for iterations prior to creating a final product. Optimization via an iterative process involves touching and feeling the prototype, in real time, in order to finalize the shape and geometry, leading to a final product. Characterization methods during iterations and on the final design include optical microscopy, SEM, and mechanical tests. Others methods, such as bio-compatibility (cell-adhesion and proliferation) and electrical performance are performed depending on the application. **Figure 1** demonstrates a general scheme for the AM process. Despite the many advances in AM, the technology still has many challenges that need to be addressed. These challenges are related to the speed of the processes (which in many cases is slower than injection molding processes and machining), cost of the machines, and limited feedstock. However, advantages outweigh the challenges due to the fact that AM allows for compositional flexibility, complex macro and microstructures, and easy modeling and optimization. As a result, industries including biomedical engineering, transportation, and the military have adopted AM as the main manufacturing method for the printing of prototypes and final parts [16, 17].

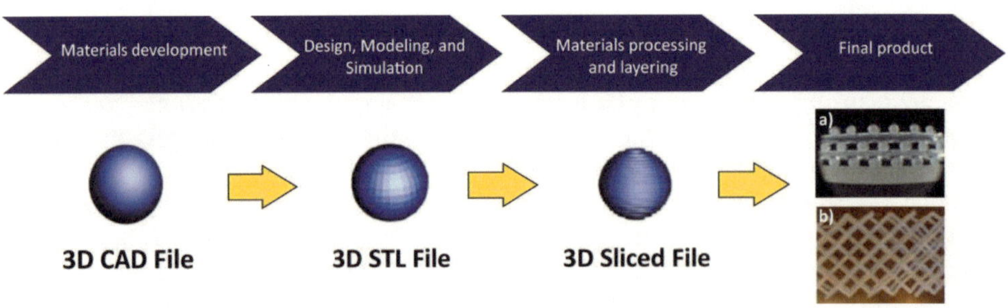

Figure 1.
General scheme for the use of additive manufacturing processes, from the choice of material to the final product. The 3D printing of parts involves the use of a computer-assisted design software that generates a STL file format that is then sliced and formatted into gcode. The computer controls the stage and dispenser to generate materials with specific architectures, e.g. faced-centered tetragonal cushion using direct ink writing (a) and diamond structure using FDM (b).

3. Polymers in 3D printing industry

Careful attention is imperative when choosing a material to print a given part. While there are a variety of commercially available polymers, not one polymer is inclusive and will give one the properties needed for a specific application. Furthermore, a single AM technique is not capable of printing any one individual polymer available in the market. The selection of material depends on the application and the customers' needs. **Figure 2** lists the decision criteria for the selection of a material. One must take into consideration the environment at which the part will be exposed and the properties required (e.g. temperature, mechanical load, humidity, chemical exposure, radiation, UV light), the processability, 3D printing method, and availability.

Polymers have become consumer goods, for they are used to manufacture bottles, toys, tools, bags, phones, computers, tools, cushions, electronics and transportation components [18]. Thus, it makes sense that efforts have focused on developing materials that can be 3D printed, which allows for rapid manufacturing [2–4, 17]. **Table 1** lists commercially available polymers used in some of the AM processes. Polycarbonate (PC), acrylonitrile butadiene styrene (ABS), poly ether ester ketone (PEEK), polyetherimide (ULTEM) and Nylon are common polymers used in processes requiring thermoplastics, or plastics that are processed by heating to a semi-liquid state and close to the melting point. Upon extrusion, the printed layers fuse and solidify. AM techniques that use thermoplastics are Fused-Deposition Modeling (FDM), Jetting (InkJet), and Selective Laser Sintering (SLS). SLA and Direct Ink Writing (DIW) use thermosetting polymers in their liquid state, or polymers that become solids after curing. A chemical reaction occurs prior to the melting point, resulting in a solid-state material. In SLA and DIW, polymers are formulated to meet specific properties, most importantly rheological. For example, each layer should be self-supporting and should allow for the printing of multiple layers while retaining the designed geometry [14, 19–21]. Rheologically, this corresponds to a resin that has a yield stress at high oscillatory stresses, such that the

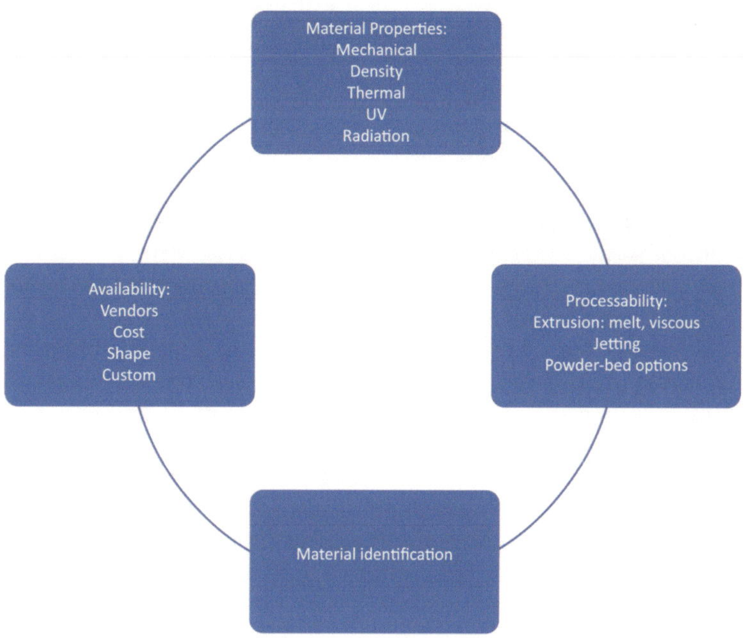

Figure 2.
Material selection chart for product design and manufacturing.

AM technology	Process	Physical state of starting material	Feedstock
FDM	Melting-solidifying	Solid	PC, ABS, PLA, ULTEM, Nylon, Carbon-filled Nylon, ASA
SLA	Photocuring	Liquid	Thermosetting- acrylates and epoxy
SLS	Melting-solidifying	Solid	PCL, PLA
Jetting	Photocuring	Solid	ABS, ASA, PCL, PLA, Vero
Direct Writing	Extrusion-heat/ UV curing	liquid	Thermosetting- any material with adequate viscosity

Table 1.
List of polymers used for 3D printing applications.

resin is solid-like at rest (low stress) and liquid like during flow (high stress) [7]. One of the main challenges in the polymer 3D printing industry is the limited feedstock available for purchase. Polymers listed in **Table 1** cannot be used in all applications. Particularly, polymers in the pure state lack mechanical strength for load-bearing applications. The addition of fillers, such as silica [22, 23] and carbon fibers [24, 25], is often used to generate materials with high mechanical strength. Furthermore, the incorporation of additives enhances materials properties by adding functionality to the parts that include getter [20], UV and radiation resistance [26–28], and anti-fouling properties [29–31].

4. Biomedical engineering applications for 3D printing

The biomedical market represents 11% of the total AM market share today, and will be a strong driver for AM development and growth [32]. Since the early 2000s, there has been increased interest in using 3D printing to fabricate hard tissues (bones, teeth, cartilage) and soft tissues (organs, skin, and others) [2–4, 16, 33]. The manufacturing of prostheses and scaffolds with complex geometries is especially important for regenerative medicine, where a porous scaffold is implanted into the patient to serve as a template for tissue to regenerate while the implant degrades slowly in the body. Other implants need to stay in place for the lifetime of the patient. 3D printing allows for the rapid manufacturing of customized prosthetics and implants with controlled architectures. The structure can be designed through the translation of x-ray, MRI, and CT images into STL file formats. The STL file can be processed by software and a design can be generated based on the patient's specific needs. Metals are commonly used to generate prosthetics for bone reconstruction. ABS and PLA are the most suitable non-biodegradable polymers used for the manufacturing of scaffolds. However, materials used in medicine must enable cell adhesion, growth, and differentiation. Current feedstock for biomaterials is limited to collagen, gelatin, fibrin, and chitosan, which are similar to natural tissue, have high affinity to cells and are highly hydrated. The main challenge with these soft natural polymers is their low mechanical strength [33]. In biomedical engineering, the main focus has been on the development of biopolymeric materials for tissue and scaffold generations with improved flexibility, strength, and patient compatibility in order to prevent implant rejection and toxicity. Some polymeric mixtures include living cells isolated from the patient and grown in the laboratory. These types of polymers are often hydrogels suitable for ink jet 3D printing technologies. **Table 2** shows various polymers used for biomedical applications.

Material	3D printing techniques	Comments
PLA, PCLA, PLGA	FDM	Scaffolds. Biodegradable. Can add fillers, e.g. HA, for improved cell adhesion and mechanical properties
Collagen, alginate, PEG, fibrin, chitosan	Inkjet, extrusion	Biodegradable scaffolds. Can add fillers and cells for improved cell adhesion and mechanical properties
PCL, methacrylate copolymers	SLS	Biodegradable scaffolds. Improved mechanical properties

Table 2.
Polymers and processes used for the additive manufacturing of biomedical devices.

Some examples of biomedical devices developed using 3D printing are implants, prosthetics, dental, orthodontics, hearing aids, and drug release tissues.

Polymers used for tissue and organ fabrication need to have various functions in order to (1) allow for cell attachment and migration, (2) transfer growth factors and waste products, (3) maintain its shape while cells are growing and (4) maintain adequate mechanical properties. Wu et al. [34] reported the generation of a biopolymeric material based on chitosan dissolved in an acid mixture of acetic acid, lactic acid, and citric acid. This biomaterial was 3D printed using an ink-writing technique, then dried under vacuum and neutralized to remove any acid residue. The structure of the scaffold was characterized using confocal laser scanning microscopy and the images showed wrinkles attributed to the volume change. Tensile mechanical tests show that the printed material exhibits a strain to failure of 400% under tensile load and a 7.5 MPa ultimate strength when in its neutralized form. Furthermore, the 3D printed material allows for excellent cell adhesion, growth, and proliferation, as demonstrated using the Live-Dead staining method, fluorescence microscopy, and SEM.

Luo et al. [35] reported the 3D printing of a bioceramic hollow struts-packed scaffold using an extrusion typ. 3D printer and a shell/core nozzle. The ink contained $Ca_7Si_2P_2O_{16}$, alginate and Pluronic F-127. After printing, the ink was dried overnight and sintered for 3 hours at 1400°C to remove the alginate and F-127 materials. The morphology was analyzed using an optical microscope. The micropores and the microstructure of the pores were characterized using SEM. The fabricated scaffolds (16/23 shell/core size) were subjected to mechanical testing and exhibited a compressive strength of 5 MPa, comparable to cancellous bone (2–12 MPa), and a modulus of 160 MPa. The scaffold had high porosity (65–85%), adjusted with the core/shell size nozzles. The high porosity and surface area (up to 6500 mm^2/g) allowed for cell adhesion and proliferation on the outer and inner surface of the scaffold, as determined by SEM. Finally, the in-vivo bone formation study in a rabbit demonstrated that the bioceramic implant allows for good cell integration and bone formation was detected with micro-CT.

Lewis' team at Harvard University 3D printed a tympanic membrane scaffold composed of PDMS, PLA, and PCL based materials using a DIW technique [36]. The team demonstrated that it is possible to design and fabricate materials with similar properties when compared to human specimens. The high frequency displacement and acoustics were organized by concentric rings for each 3D printed graft, and it was very dependent on the patterns and mechanical properties, characterized via digital opto-electronic holography, laser Doppler vibrometry, and dynamic mechanical analysis. In a different study, the team 3D printed cellular materials with vascular networks for flow [37]. The 3D printed structure was fabricated using an ink composed of Pluronic F-127, GelMA (gelatin methacrylate to allow for UV curing) and fibroblast cell culture. After curing, the Pluronic F-127 was removed by cooling to 4°C, yielding open channels that represent the vascular networks. Lewis' team demonstrated that blood and other cellular liquids can flow through the channels with minimal death of cells.

Patients with skin burns and thick wound injuries often suffer from long term recovery and extensive and expensive treatments. The autologous split-thickness skin graft (ASSG) is the technique most often used to treat large wounds [38]. A skin tissue is place in the injured area and assists with the wound closure and healing. This technique relies on the removal of a piece of skin from a different part of the patient's body and reapplying it on the place of injury. The drawback with ASSG is that it is limited by the size of donor sites and also creates another place of injury [38]. 3D printing of biomaterials would alleviate the problems related to ASSG. Skin cells are cultured in a laboratory and mixed with biocompatible polymers for bioprinting. In 2012, Koch Singh et al. [39] reported the 3D printing of skin using a laser-based inkjet printing method. The inks were composed of blood plasma/alginate solution and fibroblast/keratinocytes/collagen biomaterials. Collagen is the main component of the extracellular matrix (ECM) in skin. The team proved that the laser-based printing method does not harm the cells by performing proliferation of the cells in histologic sections 10 days after printing. Ki-67 staining, which includes the protein present in cells during their active cell cycle phases, shows that proliferating cells can be found in all regions, verifying vitality. In addition, a build-up of basal lamina, cell adhesion and proliferation- sign of tissue generation was observed.

The dental industry is taking advantage of 3D printing technologies for restoratives, implants, and orthodontics purposes. Currently, professionals in the dental field have access to 3D printers and it is possible to print designs in a clinical environment. A CT scan is used to generate a defined shape based on the patient's morphology and quickly fabricate and replace a missing tooth [40]. 3D printing is used for the manufacturing of aligners, braces, dental implants, and crowns [40]. Biocompatible materials are used for the fabrication of dental parts using 3D printing, e.g. polylactic acid, polycaprolactone and polyglycolide, and acrylates [3]. It is possible to fabricate dental implants with antibacterial properties by the incorporation of additives, such as quaternary ammonium salts [41–43]. At the age of 23, Amos Dudley fabricated his own orthodontic aligners while he was a student at New Jersey Institute of Technology [44]. He used equipment available at the institute to scan and print models of his teeth. A non-toxic plastic was used to mold and eventually generate 12 clear aligners. Amos had access to a Stratasys Dimension 1200 3D printer and used a mixture of alginate powder and PermaStone as the resin to print the aligners, which were tested by fitting them on his teeth. While it was not a trivial problem to solve, Amos proved the ability of 3D printing orthodontic materials for teeth alignment.

AM has been widely used in the biomedical industry and will continue to impact work in the future. Some challenges will persist, such as regulatory issues, limited materials, and inconsistent quality [45]. AM biomedical products require FDA approval, which can be time consuming and difficult to obtain [46]. Biocompatibility will require the development of new techniques and materials to produce high quality, high performing AM materials [47]. Furthermore, mechanical properties of AM materials need to be well assessed such that final properties can have reliable and reproducible behaviors. Further development for on-demand and patient-specific applications will be exciting work in this field. For example, designing patient-specific implants following a CT-scan will result in quick results [48]. Complex parts with specific mechanical properties and biocompatibility can be constructed on demand and with multifunctional components if needed. AM Research and development may help to improve bio-printed scaffolds and tissues for clinical applications to reduce cost for tissue engineering [49]. Manufacturing AM artificial organs, which includes multifunctionality (i.e. bionic ear [50]), will revolutionize the field of 3D printing for biomedical applications.

5. Additive manufacturing in the aerospace industry

One of the most promising fields in the future of AM is the aerospace industry. According to Wohlers' report, this industry account for almost 20% of the total AM market today [32]. Aerospace applications typically require light weight and high strength materials. The importance of AM relies on the reduced cost, increased flexibility of design, and increase in a variety of products to meet customer needs. Additive manufacturing is an important technology that enables the design and manufacturing of complex structured products with improved mechanical strength and lower weight, at a lower cost and reduced lead-time. The aerospace industry has replaced the conventional manufacturing methods of molding and machining with 3D printing technology for small scale production. At a small production scale, AM offers effectively low-cost design and assembly [17].

The aerospace industry implemented the use of AM approximately 20 years ago [51]. The main use for 3D printing has been focused on prototyping, modeling and producing jigs, fixtures and tools [17]. Furthermore, AM is used to build replacement parts on-demand when required. The ability to build on-demand spare components reduces costs for the production of parts that may never be used due to them becoming obsolete to new technology, which also saves warehouse storage space. For example, BAE Systems is currently 3D printing window breather pipes used in jetliners [52]. These pipes cost 40% less than pipes manufactured using injection molding processes and are manufactured on an as-needed basis.

Recently, NASA designed a rover, named Desert RATS, that can support humans in a pressurized cabin in space [53]. The rover is intended to transport humans to Mars. It contains 70 3D printed parts that include flame-retardant vents and housings, camera mounts, large pod doors, front bumpers, complex electronics, and others. The materials used for the 3D printing of the part used in the rover were ABS, PCABS and PC, and were printed using a FDM Stratasys 3D printer. Piper Aircraft manufactures tools using PC that can withstand hydroforming pressures of 3000 to 6000 psi. Aurora Flight Science additively manufactured wings that weigh one third of the fully dense metal components [54]. Some wings have integrated electronics. Lepron generated 200 different designs for use in piloted helicopters [17]. It is foreseen that aerospace companies will replace small components with 3D printed parts, thus reducing the weight of the machines. Some examples are arm rests, seat belts, food trays, and many others [17].

Companies have adopted AM for fast production without making substantial changes to their products [17]. This modification is mostly due to the fast-changing market and low cost of generating such small builds. Several challenges would have to be overcome to facilitate the growth of AM. Some of these challenges include: (1) current speed of AM machines is slow for bulk production; (2) few polymeric material options; and (3) current machines do not allow for the manufacturing of large components [17, 55]. In the future, it is expected that companies will pursue a completely different business model by performing product customization for end-product while maintaining the on-demand part supply. Future work will focus on the development of multifunctional structures with complex geometries, which allows for novel solutions for complicated problems. AM techniques, such as using functionally graded materials, can be used in order to tailor the mechanical and/or thermal response of components [56]. Furthermore, on-demand manufacturing will reduce costs and eliminates potential damage caused by storage [45].

6. 3D printed electronics

Electronic devices require suitable mechanical, geometrical, and optical functionalities to allow for miniaturization, low energy consumption, and smart capabilities [57]. The production of prototypes and end-products has to rapidly change due to the fast-changing technology. The conventional method for manufacturing electronic devices is using subtractive methods that involve masking and etching of sacrificial materials [58]. AM allows for the reduction of material waste, energy consumption and processing time and steps. 3D printing is being used to substitute steps for mounting and assembling electronic devices [59]. The additive process deposits material in a controlled layer-by-layer process allowing the manufacturing of complex geometries and dimensions. In addition, it enables 3D orientation of important components to improve performance. With miniaturization, AM allows for the manufacturing of small parts that would otherwise be difficult to obtain. AM has found application for thin films [60], inductors [61], solar cells [62], and others. The most common 3D printing techniques for electronics are inkjet and direct writing of conductive inks.

Jennifer Lewis and colleagues fully 3D printed a quantum-dot (QD) light-emitting diode (LED) system, including green and orange-red light emitters embedded in a silicone matrix [63]. The printed device exhibits a performance of 10–100-fold below the best processed QD-LEP but could potentially be optimized with the addition of an electron-transport layer. A copper nanoparticle stabilized with polyvinyl pyrrolidine was mixed with 2-(2-butoxyethoxy)ethanol to prepare ink for inkjet printing [64]. The ink was printed onto a polyimide subtracted and sintered at 200°C. The prepared electronic device resulted in low electrical resistivity ($\geq$ 3.6 μΩcm, or $\geq$ 2.2 times the resistivity of bulk copper). Bionic ears were printed using an inkjet printer [50]. The inks were composed of cell-cultured alginate and chondrocytes hydrogel matrix and a conductive polymer consisting of silicone and silver nanoparticles. The 3D printed ears exhibit enhanced auditory sensing for radio-frequency reception allowing the ear to listen to stereo music. This result demonstrates that bioengineering and electronics can be merged, resulting in advanced technologies. Students from Northwest Nazaren University and Caldwell High School designed the 3D printed CubeSat [65]. The CubeSat was launched aboard Delta II rocket as part of a NASA mission in 2013. It carries miniaturized electronics and sensors and is intended to collect real-time data on the effects of the harsh environments of space (oxygen, UV, radiation, temperature and collisions) on the polymeric materials- ABS, PLA, Nylon, and PEI/PC ULTEM.

Future research and development in the electronics field will take advantage of low cost methods, flexibility in design, and fast speed of 3D printers for designing and prototyping new products. For example, printing circuit boards will offer superior accuracy and flexibility, with potential cost savings, environmental impacts, faster production times, and increased design versatility. Furthermore, adaptive 3D printing, which takes advantage of a closed-loop method that combines real-time feedback control and DIW of functional materials to construct devices on dynamic surfaces, is an exciting field of research [66]. This method of 3D printing may lead to new forms of smart manufacturing technologies for directly printed wearable devices. New possibilities will emerge in the wearable device industry, in biological and biomedical research, and in the study and treatment of advanced medical treatments.

7. Recycled plastics for 3D printing

Unsurprisingly, the amount of plastic pollution on the planet is alarming [67]. Plastics have dominated our marketplace due to their utility and versatility and make up at least 10% by mass of our waste streams. Plastics are designed to be durable and to withstand harsh environmental conditions. Therefore, the amount of plastic waste is only expected to increase in the future. Currently, 91% of plastic is not being recycled. The negative impact plastics have on our ecosystem is well recognized and researchers are using this as a business model and opportunity [68, 69]. Considerable efforts are being placed on recycling and reusing plastic waste. Prof. Sahajwalla at the University of New South Wales Sydney and her team work on turning plastic waste into usable polymers, including 3D printing polymers [70]. The company Reflow is collecting polyethylene terephthalate (PET) waste bottles and turning them into filaments suitable for 3D FDM printers [71]. A company in Belgium, Yuma, is using recycled plastics for the 3D printing of sunglasses [72]. The U.S. Army Research Laboratory and the U.S. Marine Corps are working together to repurpose plastic waste by printing items from recycled plastic useful for soldiers [73]. This process allows for a decrease in transportation costs and manufacturing of parts on demand. This large effort is expected to have a positive impact on both the environment and communities by turning polymer 3D printing into income for waste collectors and removing waste from the streams.

8. Conclusions

Industries are moving toward the implementation of 3D printing as a manufacturing process because it facilitates the design of complex structures and rapid production of prototypes. AM utilizes a computer-aided design software that allows for the design of architectures with defined porosity and structures at a microscopic level. Because of the easy production of 3D printed prototypes, modeling based on a specific application can be performed to further improve the design of the end product and potentially reduce failure risks. The 3D printing of polymers and polymer composites has significantly progressed over the last 40 years and is expected to increase in the near future. Thermoplastic materials are readily commercially available for use in FDM, SLS, and inkjet processes. Materials like PC, ABS, PLA, ULTEM, and PCLA are commonly used for the manufacturing of tools, prototypes, and items used in the aerospace industry. However, these polymers are not one-size-fits-all types of polymers and are not necessary a good choice for all applications. Thus, research efforts are focused on developing materials that are capable of meeting specific applications. For examples, polymers blended with cultured cells can be used for scaffolds and implants on biological systems. Cells can be obtained from the patient and cultivated in the laboratory, thus producing a material that is less likely to be rejected by the patient. Fillers and additives can be used to generate multifunctional materials with improved mechanical properties. Fillers, such as CNTs and graphene, can be incorporated into the polymer to produce a material that is electrically conductive.

Despite all of the advances in the design and development of new polymeric materials for AM applications, challenges still remain. The availability of polymeric inks suitable for extreme applications, such as low temperature environments, high load pressures, and radiation resistance, is very limited. The development of new materials is necessary to increase the usefulness of polymer 3D printing technologies. Ideally, some of these composites are recyclable and/or biodegradable to reduce the negative impact plastics have on our environment.

Acknowledgements

We thank the US Department of Energy's National Nuclear Security Administration contract DE-AC-52-06NA25396 for providing financial support.

Conflict of interest

The authors declare no conflict of interest.

Author details

Denisse Ortiz-Acosta* and Tanya Moore
Los Alamos National Laboratory, Los Alamos, New Mexico, USA

*Address all correspondence to: denisse@lanl.gov

References

[1] Hull CW. Apparatus for Production of Three-Dimensional Objects by Stereolithography. USA Patent: US4575330A; 1986

[2] Wang X, Jiang M, Zhou ZW, Gou JH, Hui D. 3D printing of polymer matrix composites: A review and prospective. Composites Part B-Engineering. 2017; **110**:442-458. DOI: 10.1016/j.compositesb.2016.11.034

[3] Ligon SC, Liska R, Stampfl J, Gurr M, Mulhaupt R. Polymers for 3D printing and customized additive manufacturing. Chemical Reviews. 2017;**117**: 10212-10290. DOI: 10.1021/acs.chemrev.7b00074

[4] Stansbury JW, Idacavage MJ. 3D printing with polymers: Challenges among expanding options and opportunities. Dental Materials. 2016; **32**:54-64. DOI: 10.1016/j.dental.2015.09.018

[5] Ford S, Despeisse M. Additive manufacturing and sustainability: An exploratory study of the advantages and challenges. Journal of Cleaner Production. 2016;**137**:1573-1587. DOI: 10.1016/j.jclepro.2016.04.150

[6] Lewis JA. Advanced functional materials. Advanced Functional Materials. 2006;**16**:2193-2204. DOI: 10.1002/adfm.200600434

[7] Duoss EB et al. Three-dimensional printing of elastomeric, cellular architectures with negative stiffness. Advanced Functional Materials. 2014; **24**:4905-4913. DOI: 10.1002/adfm.201400451

[8] Wong KV, Hernandez A. A review of additive manufacturing. ISRN Mechanical Engineering. 2012;**2012**: 1-10. DOI: 10.5402/2012/208760

[9] Shirazi SFS, Gharehkhani S, Mehrali M, Yarmand H, Metselaar HS, Kadri NA, et al. A review on powder-based additive manufacturing for tissue engineering: Selective laser sinering and inkjet 3D printing. Science and Technology of Advanced Materials. 2015;**16**:033502. DOI: 10.1088/1468-6996/16/3/033502

[10] de Gans B, Duineveld PC, Schubert US. Inkjet printing of polymers: State of the art and future developments. Advanced Materials. 2004;**16**:203-213. DOI: 10.1002/adma.200300385

[11] Gibson I, Rosen DW, Stucker B. 3D Printing, Rapid Prototyping, and Direct Digital Manufacturing. New York: Springer; 2010. ISBN: 978-1-4939-2113-3

[12] Gross BC, Erkal JL, Lockwood SY, Chen CP, Spence DM. Evaluation of 3D printing and its potential impact on biotechnology and the chemical sciences. Analytical Chemistry. 2014;**86**: 3240-3253. DOI: 10.1021/ac403397r

[13] Turner BN, Strong R, Gold SA. A review of melt extrusion additive manufacturing processes: 1. Process design and modeling. Rapid Prototyping Journal. 2014;**20**:192-204. DOI: 10.1108/RPJ-01-2013-0012

[14] Jungst T, Smolan W, Schacht K, Scheibel T, Groll J. Strategies and molecular design criteria for 3D printable hydrogels. Chemical Reviews. 2016;**116**:1496-1539. DOI: 10.1021/acs.chemrev.5b00303

[15] Lewis JA, Gratson GM. Direct writing in three dimensions. Materials Today. 2004;7:32-39. ISSN: 1369 7021

[16] Ventola CL. Medical applications for 3D printing current and projected uses. Pharmacy and Therapeutics. 2014;**39**: 704-711

[17] Coykendall J, Cotteleer M, Holdowsky J, Mahto M. 3D Opportunity in Aerospace and Defense. Deloitte University Press, USA. DUPress.com

[18] Song JH, Murphy RJ, Narayan R, Davies GB. Biodegradable and compostable alternatives to conventional plastics. Philosophical Transactions of the Royal Society, B: Biological Sciences. 2009;**364**:2127-2139. DOI: 10.1098/rstb.2008.0289

[19] Maiti A, Small W, Lewicki JP, Weisgraber TH, Duoss EB, Chinn SC, et al. 3D printed cellular solid outperforms traditional stochastic foam in long-term mechanical response. Science Reports Nature. 2016;**6**:26573. DOI: 10.1038/srep24871

[20] Ortiz-Acosta D, Moore T, Safarik DJ, Hubbard KM, Janicke M. 3D-printed silicone materials with hydrogen getter capability. Advanced Functional Materials. 2018;**28**:1707285. DOI: 10.1002/adfm.201707285

[21] Schmalzer AM, Cady CM, Geller D, Ortiz-Acosta D, Zocco A, Stull J, et al. Gamma radiation effects on Siloxane-based additive manufactured structures. Radiation Physics and Chemistry. 2017; **130**:103-111. DOI: 10.1016/j.radphyschem. 2016.07.020

[22] Qu L, Wang L, Xie X, Gouzhu Y, Bu S. Contribution of silica-rubber interactions on the viscoelastic behaviors of modified solution polymerized styrene butadien rubbers (M-S-SBRs) filled with silica. RSC Advances. 2014;**4**:44354-64363. DOI: 10.1039/c4ra09492a

[23] Leblanc JL. Rubber-filled interactions and rheological properties in filled compounds. Progress in Polymer Science. 2002;**27**:627-687. DOI: 10.1016/S0079-6700(01)00040-5

[24] Wu C, Lin L, SHsu H,J, Yip M, Fang W. Static and dynamic mechanical properties of polydimethylsiloxane/ carbon nanotube nanocomposites. Thin Solid Films. 2009;**517**:4895-4901. DOI: 10.1016/j.tsf.2009.03.146

[25] Tekinalp HL et al. Highly oriented carbon fiber-polymer composites via additive manufacturing. Composites Science and Technology. 2014;**105**: 144-150. DOI: 10.1016/j.compscitech. 2014.10.009

[26] Soliman YS, Basfar AA, Msalam RI. A radiochromic film based on leucomalachite green for high-dose dosimetry applications. Radiation Measurements. 2014;**62**:45-51. DOI: 10.1016/j.radmeas.2014.01.004

[27] Gafar SM, El-Ahdal MA. A new developed radiochromic film for high-dose dosimetry applications. Dyes and Pigments. 2015;**114**:273-277. DOI: 10.1016/j.dyepig.2014.11.021

[28] Gonzalez-Perez G, Burillo G. Modification of silicone sealant to improve gamma radiation resistance, by addition of protective agents. Radiation Physics and Chemistry. 2013;**90**:98-103. DOI: 10.1016/j.radphyschem. 2013.03.014

[29] Dalsin JL, Messersmith PB. Bioinspired antifouling polymers. Materials Today. 2005;**8**:38-46. DOI: 10.1016/S1369-7021(05)71079-8

[30] Magin CM, Cooper SP, Brennan AB. Non-toxic antifouling strategies. Materials Today. 2010;**13**:36-44. DOI: 10.1016/S1369-7021(10)70058-4

[31] Ngo BKD, Grunlan MA. Protein resistant polymeric materials. ACS Macro Letters. 2017;**6**:992-1000. DOI: 10.1021/acsmacrolett.7b00448

[32] Wohler T. 3D Printing and Additive Manufacturing State of the Industry Annual Worldwide Progress Report. Wohlers Report; 2017

[33] Wang XH et al. 3D bioprinting technologies for hard tissue and organ engineering. Materials. 2016;**9**:802

[34] Wu Q, Maire ML, Therriault SD, Heuzey MC. 3D printing of microstructured and stretchable chitosan hydrogel for guided cell growth. Advanced Biosystems. 2017;**1**: 1700058. DOI: 10.1002/adbi.201700058

[35] Luo Y, Zhai D, Huan Z, Zhu H, Xia L, Chang J, et al. Three-dimensional printing of hollow-struts-packed Bioceramica scaffolds for bone regeneration. ACS Applied Materials & Interfaces. 2015;**7**:24377-24383. DOI: 10.1021/acsami.5b08911

[36] Kozin ED, Black NL, Cheng JT, Cotler MJ, McKenna MJ, Lee DJ, et al. Design, fabrication, and in vitro testing of novel three-dimensionally printed tympanic membrane grafts. Hearing Research. 2016;**340**:191-203. DOI: 10.1016/j.heares.2016.03.005

[37] Suntornnond R, Tan EYS, An J, Chua CK. A highly printable and biocompatible hydrogel composite for direct printing of soft and perfusable vasculature-like structures. Scientific Reports. 2017;**7**:16902. DOI: 10.1038/s41598-017-17198-0

[38] Wax MK, Meyers AD. Split-Thickness Skin Grafts [Internet]. Medscape. 2017. Available from: https://emedicine.medscape.com/article/876290-overview. [Accessed: 2018-06-28]

[39] Koch L, Deiwick A, Schlie S, Michale S, Gruene M, Coger V, et al. Skin tissue generation by laser cell printing. Biotechnology and Bioengineering. 2012;**109**:1855-1863. DOI: 10.1002/bit.24455

[40] Gadget L. Dental 3D Printing: How Does it Impact the Dental Industry? [Internet]. Sculpteo. 2017. Available from: https://www.sculpteo.com/blog/2017/12/18/dental-3d-printing-how-does-it-impact-the-dental-industry/ [Accessed: 2018-06-28]

[41] Grischke J, Eberhard J, Stiesch M. Antimicrobial dental implant functionalization strategies–A systematic review. Dental Materials Journal. 2016;**35**:545-558. DOI: 10.4012/dmj.2015-31

[42] Ge Y, Wang SP, Zhou XD, Wang HH, Xu HHK, Cheng L. The use of quaternary ammonium to combat dental caries. Materials. 2015;**8**:3532-3549. DOI: 10.3390/ma8063532

[43] Yue J, Zhao P, Gerasimov J, van de Lagemaat M, Grotenhuis A, Rustema-Abbing M, et al. 3D-printable antimicrobial composite resins. Advanced Functional Materials. 2015;**25**: 6756-6767. DOI: 10.1002/adfm.201502384

[44] Macdonald F. A College Student has 3D-Printed his Own Braces for Less than $60 [Internet]. Science Alert. 2016. Available from: https://www.sciencealert.com/a-college-student-has-3d-printed-his-own-braces-for-less-than-60 [Accessed: 2018-06-28]

[45] Ngo TD, Kashani A, Imbalzano G, Nguyen KTQ, Hui D. Additive manufacturing (3D printing): A review of materials, methods, applications and challenges. Composites Part B: Engineering. 2018;**143**:172-196. DOI: 10.1016/j.compositesb.2018.02.012

[46] Morrison RJ, Kashlan KN, Flanangan CL, Wright JK, Green GE, Hollister SJ, et al. Regulatory considerations in the design and manufacturing of implantable 3D-printed medical devices. Clinical and Translational Science. 2015;**8**:594-600. DOI: 10.1111/cts.12315

[47] Zadpoor AA, Malda J. Additive manufacturing of biomaterials, tissues, and organs. Annals of Biomedical

Engineering. 2017;**45**:1-11. DOI: 10.1007/s10439-016-1719-y

[48] Banks J. Adding value in additive manufacturing: Researchers in the United Kingdom and Europe look to 3D printing for customization. IEEE Pulse. 2013;**4**:22-26. DOI: 10.1109/MPUL.2013.2279617

[49] Di Bella C, Duchi S, O'Connell CD, Blanchard R, Augustine C, Yue Z, et al. In situ handheld three-dimensional bioprinting for cartilage regeneration. Journal of Tissue Engineering and Regenerative Medicine. 2018;**12**:611-621. DOI: 10.1002/term.2476

[50] Mannoor MS, Jiang Z, James T, Kong YL, Malatesta KA, Soboyejop WO, et al. 3D printed bionic ears. Nano Letters. 2013;**13**:2634-2639. DOI: 10.1021/nl4007744

[51] Hiemenz J. Additive Manufacturing Trends in Aerospace [Internet]. Stratasys. Available from: http://usglobalimages.stratasys.com/Main/Secure/White%20Papers/SSYS-WP-AeroTrends-03-13-FINAL.pdf?v=635513030276305587 [Accessed: 2018-06-28]

[52] Molitch-Hou M. BAE Certifies 3D Printed Plastic Parts for Jetliners [Internet]. 3D Printing Industry. 2014. Available from: https://3dprintingindustry.com/news/bae-3d-printed-plastic-parts-jetliner-24567/ [Accessed: 2018-06-28]

[53] Stratasys. 3D Printing a Space Vehicle. Stratasys. ISO: 9001.2008

[54] Fresno D. Revolutionary "Smart Wing" Created for UAV Model Demonstrates Groundbreaking Technology [Internet]. Optomec. 2013. Available from: https://www.optomec.com/revolutionary-smart-wing-created-for-uav-model-demonstrates-groundbreaking-technology/ [Accessed: 2018-06-28]

[55] Gao W, Zhang Y, Ramanujan D, Ramani K, Chen Y, Williams CB, et al. The status, challenges, and future of additive manufacturing in engineering. Computer-Aided Design. 2015;**69**:65-89. DOI: 10.1016/j.cad.2015.04.001

[56] Cooley WG. Application of Functionally Graded Materials in Aircraft Structures [Thesis]. Air Force Institute of Technology; 2005

[57] Ota H, Emaminejad S, Gao Y, Zhao A, Wu E, Challa S, et al. Application of 3D printing for smart objects with embedded electronic sensors and systems. Advanced Materials Technologies. 2016;**1**:1600013. DOI: 10.1002/admt.201600013

[58] Tan HW, Tran T, Chua CK. A review of printed passive electronic components through fully additive manufacturing methods. Virtual and Physical Prototyping. 2016;**11**:271-288. DOI: 10.1080/17452759.2016.1217586

[59] Espalin D, Muse DW, MacDonals E, Wicker RB. 3D printing multifuntionality: Structures with electronics. The International Journal of Advanced Manufacturing Technology. 2014;**72**:963-978. DOI: 10.1007/s00170-01405717-7

[60] Okimoto H, Takenobu T, Yanagi K, Miyata Y, Shimotani H, Kataura H, et al. Tunable carbon nanotube thin-film transistors produced exclusively via inkjet printing. Advanced Materials. 2010;**22**:3981-3986. DOI: 10.1002/adma.201000889

[61] Liang W, Raymond L, Rivas J. 3-D-printed air-core inductors for high-frequency power converters. IEEE Transactions on Power Electronics. 2016;**31**:52-64. DOI: 10.1109/TPEL.2015.2441005

[62] Eom SH, Senthilarasu S, Uthirakumar P, Yoon SC, Lim J, Lee C, et al. Polymer solar cells based on

inkjet-printed PEDOT: PSS layer. Organic Electronics. 2009;**10**:536-542. DOI: 10.1016/j.orgel.2009.01.015

[63] Lewis JA, Ahn BY. Three-dimensional printed electronics. Nature. 2015;**518**:42-43. DOI: 10.1038/518042a

[64] Lee Y, Choi J, Lee KJ, Stott NE, Kim D. Large-scale synthesis of cooper nanoparticles by chemically controlled reduction for applications of inkjet-printed electronics. Nanotechnology. 2008;**19**:415604-415611. DOI: 10.1088/0957-4484/19/41/415604

[65] Davis S. Student-Designed 3D Printed CubeSat to Assist in Made in Space Material Research [Internet]. Tct MAG. 2017. Available from: https://www.tctmagazine.com/3d-printing-news/student-3d-printed-cubesat-assist-made-in-space-material-research/ [Accessed: 2018-06-28]

[66] Zhu Z, Guo SZ, Hirdler T, Eide C, Fan X, Tolar J, et al. 3D printed functional and biological materials on moving freeform surfaces. Advanced Materials. 2018;**30**:e1707495. DOI: 10.1002/adma.201707495

[67] Tullo AH. Fighting ocean plastics at the source. Chemical & Engineering News. 2018;**96**(16). ISSN: 0009-2347

[68] Scott A, Tullo AH. Europe ban some single-use plastics. Chemical and Engineering News. 2018;**96**(23). ISSN: 0009-2347

[69] Caliendo H. Startup Turns Recycled Plastic into 3D Printing Filament [Internet]. 2016. Available from: https://www.ptonline.com/blog/post/startup-turns-recycled-plastic-into-3d-printing-filament- [Accessed: 2018-06-28]

[70] Saunders S. Changing the Concept of Waste: Using Microfactories to Turn Trash into 3D Printing Filament and Metal Alloys [Internet]. 2018. Available from: https://3dprint.com/201565/smart-team-trash-microfactories/ [Accessed: 2018-06-28]

[71] Crew B. This New Device Recycles Plastic Bottles into 3D-Printing Material [Internet]. 2015. Available from: https://www.sciencealert.com/this-new-device-recycles-plastic-bottles-into-3d-printing-material [Accessed: 2018-06-28]

[72] World's First 3D Printed Sunglasses from Plastic Waste [Internet]. Available from: https://www.kickstarter.com/projects/1155776891/worlds-first-3d-printed-sunglasses-from-plastic-wa [Accessed: 2018-06-28]

[73] Scott C. US Military Researchers Water Bottle Recycling for 3D Printing in the Field [Internet]. 2018. Available from: https://3dprint.com/207266/military-research-recycling/ [Accessed: 2018-06-28]

Chapter 2

Self-Healing Polymer Composites for Structural Application

Jeetendra Kumar Banshiwal and Durgesh Nath Tripathi

Abstract

Self-healing materials are the next-generation materials for high-performance structures. To reduce the fatigue and subsequent probability of failure along with extended service life of polymer and polymer composites, the self-healing concept has great potential. Today, polymeric composites are structural matrix and prone to failure against cyclic mechanical and thermal loading. Significant degradation of polymeric structures at surficial sites can be measured by barely visible impact damage (BVID), but internal micro-cracks are not easily detectable. Various damage modes make major damage sites in composites and further lead to catastrophic failure of the structure. On-site repairing of microscopic or macroscopic damages in polymer composites is a value-added function that is offered by self-healing techniques. Different extrinsic methods including encapsulation, hollow fiber embedment, and vascular methods are preferred, and some intrinsic, dynamic bonding is created by reversible covalent networks and supramolecular interaction based on H-bonding, metal-ligand, and ionomers. This chapter is preferred on the new trends and challenges regarding the structural health monitoring of polymeric composites against external mechanical and environmental impacts and extended service life and performance by utilizing self-healing strategies.

Keywords: intrinsic and extrinsic self-healing, covalent reversible network, supramolecular interactions

1. Introduction

Polymers and fiber-reinforced polymers (FRPs) are common as structural materials due to lightweight, easy processability, and constancy against adverse environmental impacts. However, mechanical properties are associated with many variables including service time, operating temperature and pressure, molecular weight, and constitution of matrix. The long-term durability, high performance and reliability are major challenges for polymeric architecture. Limiting factors of polymer composite is relatively poor performance under impact loading due to lack of plastic deformation, which is a most prominent aspect of any vehicle component design. Low-velocity and high-velocity impacts are the critical issues for FRPs. These impacts influence on mechanical strength and stiffness along with dimensional stability. In metals, the impact energy is dissipated through elastic and plastic deformations, so structural integrity is retained intact. However, in FRPs the impact energy is dissipated in the form of damages in matrix. The impact damages in FRPs affect mainly the internal integrity as compared to superficial visible zone. On cyclic mechanical and thermal loadings, stress is applied on a matrix which released in the form of micro-cracking. Micro-cracks degrade polymeric properties inevitably and result into the failure. Single or multiple damage modes result into damage sites and are required to repair the components to continue to the service. If microcracks are untreated, they lead to larger macroscopic cracks, and finally catastrophic loss of the structure is an adverse outcome. To resist these failures, the new-generation materials having autonomic healing capability to damage repair are needed to develop. Conventionally, different lay-up repair techniques are adopted in thermoset.

Similarly, in thermoplastics which are known for their mechanical performances, service temperature, and solvent resistance, fusion bonding techniques are used to recover mechanical integrity.

However, these repair strategies are very costly, time-consuming, followed by complicated procedure, and assisted only by expert technicians. So, we required such unique functionality which intimates the damages at nascent stage and sometimes repaired the damages. Self-healing of cracks is an eminent and efficient possible solution of these issues. In healing systems, a damage incident triggers the internal processes that generate the healing response which cured the damages. The bio-systems have damage detection and subsequent prevention techniques; those are source of innovation to design such functionality by introducing the self-healing functionality into artificial materials. These techniques are demanding for manned materials and structures and assisted by the biomimetic approaches. The major concern is focused on recovery of mechanical properties of polymer composites against quasi-static fracture. Initially, Mercier developed a self-healing rubber that can reseal on puncture damage [1]. The concept of healing is progressed with developing efficient vehicles and other systems such as space suits to protect from micrometeoroid impacts [2]. Currently, damage healing of polymer structure is being demonstrated via different approaches, which include extrinsic and intrinsic methodologies. Extrinsic methodology is being performed via encapsulation [3–8], hollow fibers [9–11], microvascular systems [12–16] and intrinsic damage healing is offered by reversible covalent bonding [17–21] and supramolecular interactions [22–30]. In extrinsic methodology, healing is restricted only once, and the delivery of healing precursor's amount is limited. To overcome the above concern, different intrinsic healing systems are developed that offer multiple healing at macroscopic damages. Multiple healing of the same crack is achieved by microvascular 3D system, thermoreversible networks, and supramolecular interaction. These strategies help to design various self-healing systems. In literature, various high-performance systems such as self-healing coating, self-healing ceramics, and self-healing metals are reported. Self-healing nanocomposites are also reported. Self-healing technology provides public safety and reduced maintenance cost of the structure. Healing approaches offer longer lasting with fault-tolerant components across various fields including coatings, electronics, robotics, transportations, energy, etc. In the following section, the author is mainly focused to elaborate current trends and the leading research field of remendable polymers for structural composites.

2. The basic principle and theory of damage healing

Mimicking biosystems, in synthetic materials, damage repair is performed by three steps including actuation of healing, later transport of healing precursors to the damage site, and finally chemical repair process occurring with contact of catalyst or curing species at an angstrom level in which matrix is restructured by bonding of cleaved bonds at the damage site. Macroscopically, healing is proceeding by two consecutive mechanics; these included physical and chemical process. In physical process, the flow of healing precursors or segmental movement of chains is mandatory for damage repair. These events proceed continuously and are controlled by kinetics and thermodynamics. The repair event is determined by the kinetic energy of chains and entropy changes meanwhile by chain diffusions. These intrinsic properties have a great impact on entropy contribution to Gibbs free energy during repair event [31]. Another considerable factor is a free volume in a matrix which is desirable to the mobility of polymeric chains [32]. To obtain free volume, some stimuli-responsive units are also incorporated into nonreversible systems by copolymerization, and during mechanical stress, the entropy ΔS increased due to segmental mobility, and finally rebonding of cleaved sites is possible [33]. During self-healing process, voids/free volume facilitated segmental mobility of chains and matrix macromolecular chains. Void-less system are rigid and subsequently damage sensitive. Heterogeneities are critical parameters which offer the design of self-repair concept. Different types of polymers like block, branched, and/or star polymers showed self-repairs in range of nm to μm. Microphase separation and microcapsules and inorganic particles are embedded into polymers and responsible for macrodamages. In polymeric systems, heterogeneities are developed by phase separation utilized by copolymers or composite materials [27, 28, 34] and shape memory polymers [35]. In chemical process, different polymerization reactions of healing precursor or entanglement of polymer chains or reversible covalent bonding according to base matrix materials is dominant. All these stages are balanced by the damage rates to healing rate. The rate of damage is defined by various factors such as loading frequencies, strain rate, and the amplitude of stress. However, healing rate can be monitored by concentrations of precursor species and/or intrinsic temperature using varying reaction kinetics. In thermoset, encapsulation is an effective healing strategy, but in thermoplastics induced-healing is reported in which healing is possible on heating of polymers above its glass transition temperature (Tg) or using solvents by depression of the effective Tg as compared to below room temperature [36]. In autonomic healing, the healing agent is incorporated or phase separated by matrix so that the healing of crack/failure takes place without external intervention at ambient temperature. It is fully self-contained and responds to external stimuli. Healing is achieved by one-capsule system, dual-capsule system, and hollow fiber and vascular network-based system. In non-autonomic healing, human intervention is mandatory. It is inherent and intrinsically similar to biological structures. It is a partially self-contained healing system. Healing functionality is an intrinsic part of base matrix, but additional heat or radiation is required to proceed. Generally, for high healing efficiency, the healing agent forms a homogeneous mixture although it is difficult to process in terms of large-scale production in industries. The healing efficiency represents the recovery of mechanical integrity of components. To quantify healing efficiency, many definitions have been proposed. Basically, healing efficiency (η) is a ratio of change in a property of interest of materials as shown in equation no. 1.

$$\eta = \frac{f_{healed} - f_{damaged}}{f_{virgin} - f_{damaged}} \tag{1}$$

where f is the property of interest of material, and f_{healed}, $f_{damaged}$, and f_{virgin} are the property of interest of healed sample, the damaged sample, and the virgin sample, respectively.

Damage volume is a deciding factor of maximum healing efficiency of various repair systems. Each technique demonstrated diverse healing efficiency for different damage volume. Intrinsic systems are preferred to heal small damage volume and heal at molecular level due to the close proximity of damaged site which is mandatory for re-bonding of the cleaved site. Microvascular network heals large damage volume and potentially attempts highest healing efficiency. Encapsulation strategy covered the regime between intrinsic and vascular systems. Most of the repair systems established high damage volume to low healing rate. To achieve high healing efficiency, the damage rate should be equal to damage healing. Only some systems based on capsule and intrinsic system are matched to healing rate to damage rate. The exact nature of the self-healing method to be deployed depends upon (i) the nature and location of the damage, (ii) the choice of repair resin, (iii) the influence of the operational environment, and (iv) proximity of damage site and healing precursor container. The stability and durability of the final material can be increased by repairing the damage in an autonomic way. Currently, a more dynamic strategy based on damage acceptance and management has been explored and growing exponentially shown in **Figure 1**.

Multiple healing is possible through the intrinsic approaches which have intrinsic functionality. This approach can be practical to thermoplastic, thermoset polymers and elastomers. Intrinsic self-healing is achieved by the recovery of the former interactions, with or without an external trigger. A certain magnitude of stress (i.e., chemical, mechanical, or thermal) enhanced the mobility of polymer network. On impact, the sudden drop of viscosity in matrix occurs due to transfer of impact energy in the form of heat to localized zone, which allows the local deformation and mobility of polymer chains or network. Upon cooling, network restores the initial values of viscosity, and materials achieved virgin mechanical and thermal stability. Moreover, to increase mechanical properties of intrinsic system, more than one chemical healing principle may be required to combine. The damage interfaces disappear when chain entanglements and chemical or physical cross-links formed a network as strong as the bulk material. This process can be obtained by physical and chemical interactions and a combination thereof. The most accepted theory leading to interfacial physical healing is proposed by Wool and O'Connor [37] which is based on molecular interdiffusion leading to chain entanglements. This process can occur at higher temperatures as compared to the bulk polymer glass transition temperature or through local external trigger such as a solvent and temperature beyond the melting point in thermoplastics (welding) [38]. In the case of reversible chemistries, the enhanced mobility leads to a viscous flow of the material in the vicinity of the damage site. It is remarkable that chain interdiffusion has been observed also at temperatures theoretically below the bulk Tg which highlights the potential difference in Tg between the bulk and the surface in freshly damaged materials influencing the healing process. From a mechanical and theoretical point of view, up to 100% healing of an

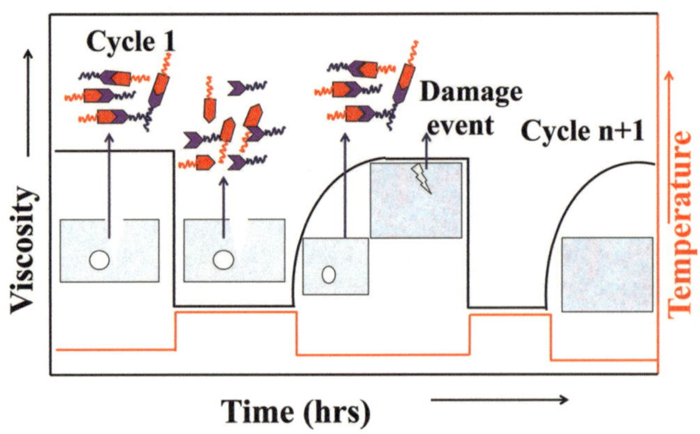

Figure 1.
The concept of damage healing using intrinsic methodology.

interface can only be obtained if the new interface has exactly the same properties as the bulk material.

3. Passive self-healing based on extrinsic techniques

In passive mode, healing is generated by incorporation of foreign functionalities. The extrinsic healing process is based on the use of a healing agent contained in the matrix as a separate phase. The healing agent is usually in the liquid state, placed into reservoirs which may be microcapsules or hollow fibers or microvascular network. In most approaches, the healing agent is used with a catalyst, which can also be encapsulated or dissolved in the matrix. Different extrinsic healing approaches are explored. In some cases, the catalyst is not required to initiate the healing process; the healing agent can also react to itself. The extrinsic healing concept is based on the response after or at the onset of damage. Current research is concerned with the improvement of healing agents in terms of compatibility and catalyst-free system with some new encapsulation techniques.

3.1 Microcapsule embedment

Encapsulation strategy is mainly studied for polymers and coating. The basic principle of strategy is healing by incorporated healing functionality or reactive constituents into capsules followed by chemical reactions. These reactions take place by various mechanisms including ring opening metathesis polymerization (ROMP) [3], cycloreversion [39], cycloaddition [40], cross-linking reactions [41], or a mechanochemical catalytic activation [42]. Damage acts as a stimulus to initiate the healing process. Damages rupture the microcapsule, and subsequent release of the core material (healing agent) is possible. The healing precursor reached at the damage site by capillary action and spreads itself over the two fracture surfaces due to the surface tension. Further, precursors interact with embedded adjacent catalysts (**Figure 3**) leading to a network formation by following the above chemistries, which terminate the further growth of crack or damage and restore mechanical integrity. White et al. [3] designed a "dicyclopentadiene (DCPD) Grubbs' system" based on capsule healing which achieved 75% recovery of virgin fracture toughness of TDCB specimens. Capsule- and hollow fiber-based healing systems are shown in **Figure 2**.

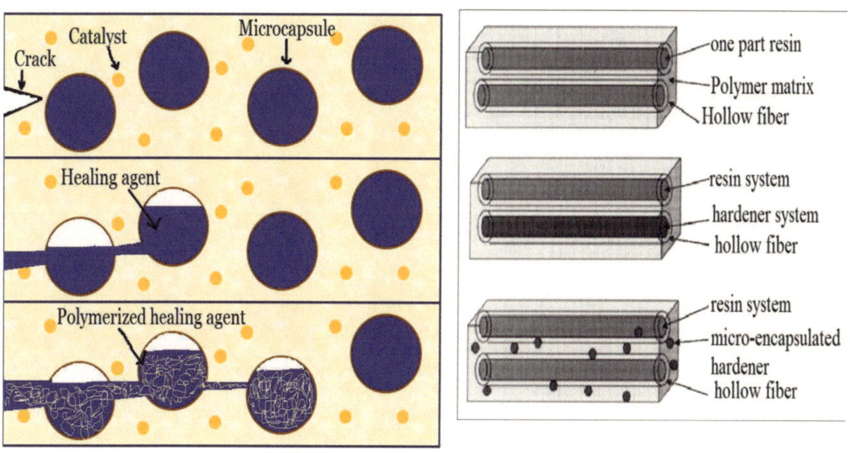

Figure 2.
Demonstration of healing phenomenon. (a) Capsule-based healing [3] and (b) hollow fiber embedment [11].

Mainly, the adhesive and cohesive mechanisms are responsible for the failure of interfaces. Better healing efficiency can be achieved by improving the adhesive tendency of the poly-(DCPD) with fracture surface of the matrix without compromise with the cohesive strength of the poly-(DCPD) (shown in **Figure 4**). The average diameter of the microcapsule is ranging from ~300 to ~700 μm and shell walls with thicknesses from 5 to 20 μm. Generally, the core materials of microcapsules are made of poly(urea-formaldehyde) (PUF) [43]. The controlled release of the precursor is the typical job of encapsulation strategy. Some of the eco-friendly catalyst-free healing methods are also introduced because the catalyst-based approach is cost-ineffective and related to some compromise of mechanical properties of matrix. Various critical factors define the performance of encapsulation strategy [44, 45]. There are various different ways to proceed capsule healing; that is, (i) the encapsulated liquid agent can be combined with a dispersed catalyst, (ii) both the healing agent and the catalyst can be embedded in different capsules, (iii) the healing agent can also directly react with a functionality of the matrix under an external stimulus, and (iv) the healing agent and the catalyst can be placed in the matrix as a separate phase. Different capsule-based healing systems are summarized in **Table 1**.

The encapsulation strategy is mainly focused on meltable dispersion and in situ and interfacial encapsulation techniques for capsules. Meltable dispersion is the method of dispersing the healing agent in a melted polymer to form the capsules after solidification of the polymer [47]. In situ and interfacial techniques have been used for PUF or TETA microcapsules. In this technique, the shell is developed by polymerization at the interface of healing agent droplets and the oil-in-water emulsion.

The triggering mechanism of encapsulation strategy is validated by optical microscopy of a fracture plane showing ruptured capsules, by infrared spectroscopy (IR), scanning electron microscopy (SEM), and energy-dispersive X-ray spectroscopy (EDS) of the fracture plane. Different microencapsulation methods are optimized such as physical methods, chemical methods, and physicochemical methods. Robust, in terms of mechanically and thermally stable, microcapsules having healing precursor have been synthesized for self-healing polyurethane matrix [48]. Triethylenetetramine (TETA) microcapsules for wear-resistant polymer composites [49] and poly(methyl methacrylate) microcapsules with high storage and thermal stability [50] have been manufactured and implemented. The switching behavior of microcapsule geometry between dry and wet condition is a critical healing phenomenon. Polydimethylsiloxane-based self-healing elastomers

S.N.	Healable systems	Healing mechanism	Ref. (s)	Healing efficiency (%)
1	DCPD/PUF-microcapsules/Grubbs catalyst	ROMP	[3, 43]	99
2	Epoxides/amine	Curing mechanism	[4]	91
3	Epoxy/DCPD-WCl_6 catalyst	ROMP	[5]	20
4	Epoxy/mercaptan	Curing mechanism	[6]	>100
5	Polysiloxane (polydimethylsiloxane)/Pt catalyst	Polycondensation reaction	[7]	>100
6	Thiol/maleimide	Michal addition reaction	[46]	>100
7	Woven glass fabric/epoxy laminates	Epoxy/ $CuBr_2$(2-methylimidazole)$_4$	[8]	68–79

Table 1.
Various self-healing systems based upon encapsulation strategy.

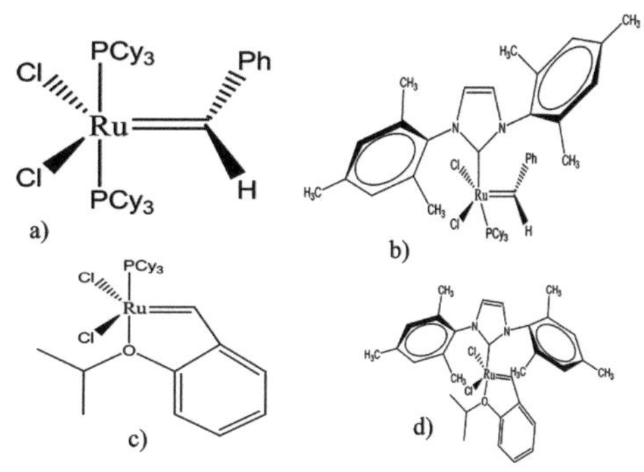

Figure 3.
(a) Grubbs first generation, (b) Grubbs second generation, (c) Hoveyda-Grubbs first generation, and (d) Hoveyda-Grubbs second generation.

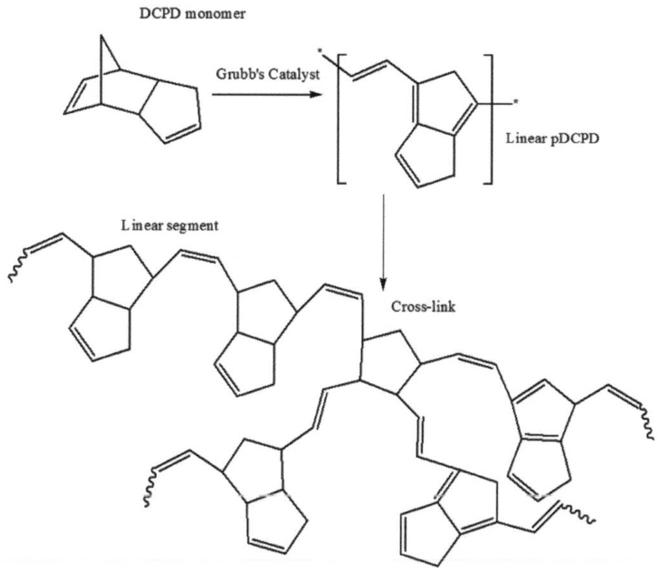

Figure 4.
The ROMP of encapsulated DCPD by Grubbs catalyst [3].

also are reported [7]. Dual component self-healing epoxy system containing epoxy (DGEBA) and different variants of hardener microcapsules are investigated [51]. Various catalysts for ROMP of DCPD are shown in **Figure 3**.

Encapsulation techniques offered single healing due to unavailability of healing precursor into capsules which is earlier invested in damage repair. However, these are limited by processing difficulties and inhomogeneous distributions of two components. Deteriorates of some mechanical integrity is common due to the addition of external chemical constituent's limiting the strategy. To deliver a larger amount of healing agent, hollow glass fibers were used. This fiber reinforcement is based on the bleeding ability of bio-system. For polymer composite systems, the hollow fiber embedment approach has been more appropriate (**Figure 4**).

3.2 Hollow fiber embedment

Hollow fibers are used to deliver a larger amount of liquid healing agent. These are embedded within either glass fiber-reinforced plastic (GFRP) or carbon fiber-reinforced plastic (CFRP) composites. Healing-agent-filled hollow fibers are introduced into the matrix by the vacuum-assisted resin transfer molding (VARTM) process. Vascular self-healing materials have appropriate healing agent in a network in the form of capillaries or hollow channels, which may be interconnected one dimensionally (1D), two dimensionally (2D), or three dimensionally (3D), upon damage. One-dimensional system is designed by glass pipettes that are embedded in epoxy resins [52]. Resin-filled hollow glass fibers impart healing capability on low-velocity impact damage in CFRP [9]. Large-diameter capillaries are not feasible to demonstrate damage healing. Smaller hollow glass fibers filled with resin have been also used, but they were unable to deliver the resin into the crack due to the high-viscous epoxy resins. Later, borosilicate hollow glass fibers (with diameters from 30 to 100 μm with 55% of hollowness) were produced to store the healing precursor resin. This approach offered certain advantages, such as the higher volume of healing agent to deliver, performed by different activation methods; visual inspection of the damaged site is possible, and embedment of hollow fibers to conventional reinforcing fibers is easier. The fracture of hollow fibers is mandatory to release healing precursors which limited the approach. The low viscosity of healing agent is favorable to facilitate fiber infiltration, which is necessary. The reinforcement of hollow glass fibers into CFRP also affects the coefficient of thermal expansion so that multistep fabrication stages of the hollow fiber are another challenge. A novel hybrid multi-scale carbon fiber/epoxy composite reinforced with self-healing core-shell nanofibers at interfaces has been demonstrated [10]. The ultrathin self-healing fibers were fabricated by means of co-electrospinning, in which liquid DCPD as the healing agent was enwrapped into polyacrylonitrile (PAN) to form core-shell DCPD/PAN nanofibers. To enhance the healing efficiency, vascular method is adopted in which a 3D microvascular network is developed into the matrix to store the healing agents for transport in longer distance.

3.3 The microvascular embedment

A microvascular technique is inspired from the respiratory system of livings. The incorporation of micro-channels with a diameter ranging from 1 μm to 1 mm within a polymer composite offers multiple healing. Self-healing materials that use hollow fibers or a mesoporous network are called vascular materials. Microvascular fabrication is possible by various techniques including laser micromachining, soft lithography, electrostatic discharge, fugitive inks, and hollow glass fibers. Healing precursors have been introduced into these channels either by pumping or through capillary forces. However, hollow glass fibers are restricted to the 1D network, but in order to obtain 2D and 3D interconnected networks, steel wires of ca. 0.5 mm could be used [53]. For the interface between plies in laminated composites, two-dimensional networks are suitable. Microvascular channels into polymer composites offer the benefits of added functionality and increased autonomy (i.e., the ability to distribute active healing material for crack healing which is difficult by conventional methods into monolithic materials). The introduction of sacrificial fibers into woven preforms enables the continuous fabrication of 3D microvascular composites that are both strong and multifunctional [12, 13]. To employ damage healing in microvascular systems, functional fluids that act as a healing agent are released upon fracture of vascular network. Further, the healing agent polymerized with adjacent catalyst formed a network and restricts the growth of damage. Active cooling microvascular systems continuously circulate a fluid into, through, and out of the matrix in order to absorb and remove excess heat. Replacement of some reinforcement fibers of FRPs by individual hollow fibers is a well-known method to achieve microvascular composites [54]. Schematic diagram of self-healing materials

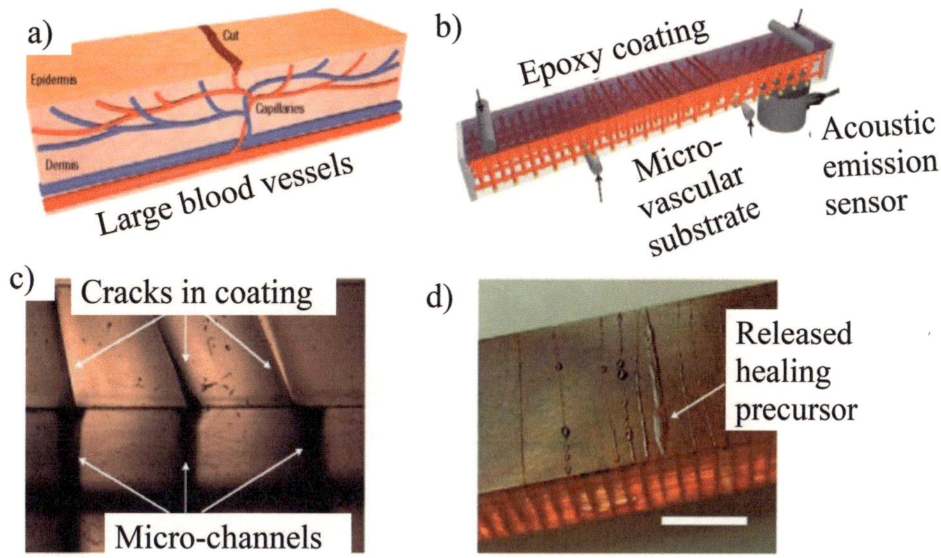

Figure 5.
Schematic diagram of self-healing materials with 3D microvascular networks. (a) Capillary network in skin having a cut in the epidermis layer. (b) Vascular network into epoxy coating having catalyst in a four-point bending configuration monitored with an acoustic emission sensor. (c) Crack propagation toward micro-channel. (d) Optical image of released healing agent in the coating [15].

with 3D microvascular networks is shown in **Figure 5**. The effect of optimum size and orientation of hollow fibers in microvascular architecture in the epoxy matrix has been studied [14].

A novel type of hollow fiber called "compartmented fibers" has been developed [55]. However, hollow fibers have a greater influence on the mechanical properties of composites than microcapsules. Indeed, using this type of fibers, a localized healing response can be activated. Vasculature-based healing allowed the efficient delivery of the healing agent and additionally healed a large area. The large-scale production of self-healing fiber-reinforced composites is not feasible due to complex vasculatures and lack of scalability. After the first healing, the network may be refilled for the next healing. Initially, a brittle polymer coating is applied to a more ductile polymer substrate which contains the interconnected network of micro-channels. In contrast, synthetic self-healing composites have high strength-to-weight ratios with less dynamic functionality.

Ultralow-temperature damage healing is achieved by incorporating a 3D microvascular network (**Figure 6**). Hollow vessels are used to deliver healing agents, and

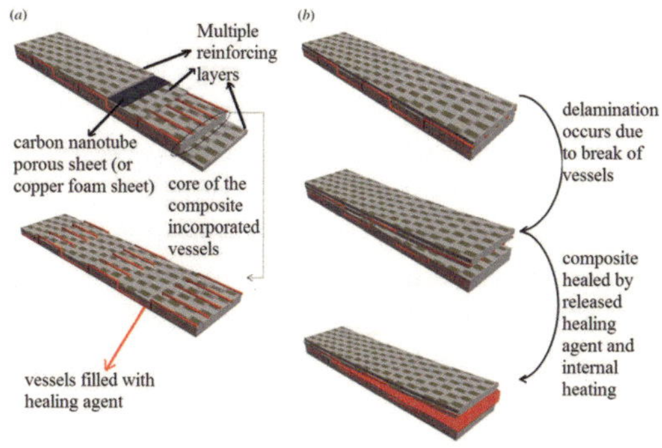

Figure 6.
(a) Internal structure of the composites and (b) damage-bleeding healing process [16].

a porous conductive wire defrosts the system by internal heating, and further healing reactions are proceeded [16]. The concept may be used to develop self-healing in aerostructure at high altitude having low temperature.

4. Active self-healing based on intrinsic methodology

Structurally dynamic polymers are a macromolecular system in which dynamic bonds are responsible for the restructuring of molecular architecture upon to external stimuli. Reversible bonding chemistry (i.e., covalent and non-covalent) is used to design structurally dynamic polymer systems produced to respond on macroscopic changes of material's architecture. Repeatable damage healing is demonstrated by various non-covalent interactions (reversible physical bonding), the covalent chemistries (reversible chemical bonding), and recently by their varied combination. Dynamic bonds are sensitive to specific stimulus and selectively undergo reversible bonding and de-bonding under equilibrium conditions. Intrinsic healing systems are programmed to respond to macroscopic damages. Basically, it is active methodology; healing is achieved by dynamic bonding of the polymer matrix. Intrinsic healing has specific properties of certain materials, such as molecular structures and chemical or physical bonds. The intrinsic self-healing requires human/external intervention to perform in most of the cases. The healing is possible via temporary local mobility enhancement of polymeric chains. Various modes of energy (e.g., temperature, static load, UV) are critical factors for the mobility of polymeric chains. Some chemical principles with multiple chemistries are responsible for macroscale healing. Reversible supramolecular interactions are low-energy interactions and based on hydrogen bonding, ionomer bonding, π-π interactions, or metal coordination. Two other categories of intrinsic healing systems combining physical and chemical approaches can be included: shape memory polymers and polymer blends.

4.1 Thermodynamic covalent bonding-based Diels-Alder (DA) and Retro-DA (r-DA) reactions

Cross-linked polymer networks have superior mechanical properties and thermal as well as chemical resistance as compared to their uncross-linked and linear analog. But, due to high cross-linked density, these systems are rigid and susceptible to mechanical damage. By incorporation of some dynamic covalent functionality into matrix backbone and/or in side chains, we achieved stimuli-responsive systems. Dynamic cross-linked systems have improved service life and energy efficiency and resist to foreign object impact. Basically, thermoreversible bonding is a more useful technique to load-bearing structures. It is accomplished by reversible chemical reactions upon an external stimulus. Typical dynamic bond chemistry is belonging to disulfide [56], hindered urea [57], and alkoxyamine [58] that are having flexible bonding units. A polymer (3M4F) system demonstrated self-repairing by subjecting it to heating/cooling cycles [18] shown in **Figure 7**.

In contrast, cycloaddition reaction is an efficient method to design carbon-carbon linkage without the use of catalyst. An electron-rich diene and electron-poor dienophile species play a key role to succeed DA/r-DA cycloaddition reaction. A product of DA reaction is known as DA adduct. DA adduct is the mixture of endo- and exo-diastereomers. Due to the temperature of the r-DA reaction, the exo-diastereomer is a major adduct. DA adducts have norbornane-type covalent functionality which is slightly weaker than other covalent linkages in matrix. Upon excess mechanical loading, the excess stress is transferred to weak bonding of adducts, and de-bonding

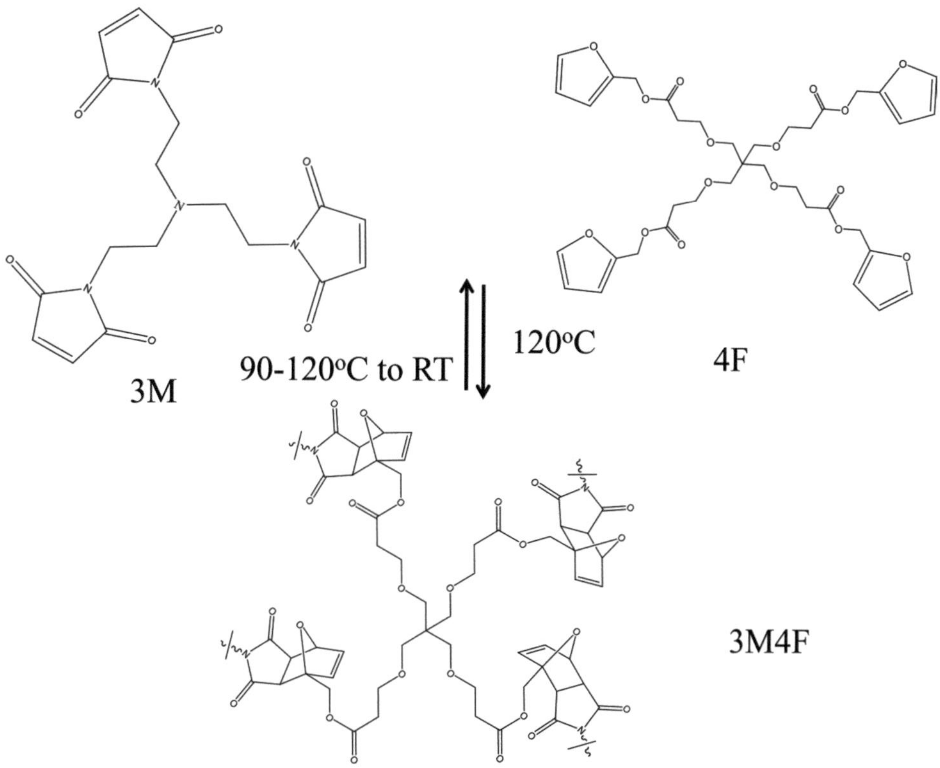

Figure 7.
Reversible cross-linked furan-maleimide-based polymer network [18].

occurs. Upon elevated temperature, DA adducts can be dissociated into corresponding diene and dienophile moieties through r-DA reaction within the cross-linked system. Upon further cooling, re-bonding is preceded into DA adduct by reaction between corresponding diene and dienophile units. Due to thermal reversibility of DA reactions, they are frequently applied to production of remendable and recyclable materials [17–20]. The 4 + 2 cycloaddition Diels-Alder reaction is the most studied thermally controlled reaction and belongs to the group of "click reactions" that are famous for flexibility and well-organized, and selective chemical synthesis [21]. By utilizing r-DA and following DA reactions, covalent network displays a dissociation and reformation through void-filling thermoreversible process upon controlled heating. Chemically, on the basis of utilized form of diene and dienophile functionalities for DA/r-DA reaction, these systems are classified into three categories: furan-maleimide polymer systems, dicyclopentadiene-based systems, and anthracene-functionalized polymer systems. Graphene nanosheet-functionalized polyurethane-based composite has shown infrared (IR) laser-assisted self-healing, which is advantageous to flexible electronics [59]. Most of the DA cross-linked network is fabricated through step-growth poly-addition or coupling reactions of polyfurans and polymaleimides. Furan and maleimide pair is highly reactive for cycloaddition reaction and low-temperature shifting of DA/r-DA equilibrium because these moieties exist in s-cis conformation, which offered rigid system favorable for DA reaction. Furan group acts as diene, and maleimide acts as dienophile based on DA/r-DA reaction chemistry. These reactions still need an external heat source to initiate the healing process. Various healing systems based on reversible covalent bonding are presented in **Table 2**.

If the amount of damage is microscopic, capsule-based or intrinsic systems may be the best option. But, macroscopic damaged volume and vascular-based systems are

S.N.	Healable systems	Healing mechanism	Ref.	Remarks
1	Dicyclopentadiene-based polymers	DA/r-DA	[17]	Polymer side chains functionalized
2	3M4F polymer	DA/r-DA	[18]	Multivalent star-shaped systems
3	2ME4F polymer	4 + 2 DA reaction	[20]	Solvent-free synthesis
4	Anthracene-based polymers	DA/r-DA	[19]	Polymer backbone functionalized along itself
5	Thiol-modified poly [N-acetylethylene-imine]	Redox-reversible hydrogel system	[60]	Thiol-disulfide system (side-chain modification)
6	Polystyrene-based block copolymer	Thiol-disulfide linkage	[61]	Thiol-disulfide system (backbone modification)

Table 2.
Various self-healing systems based upon reversible covalent bonding.

efficient which allow large amounts of healing agent to be transported to the damage site. The aforementioned self-healing techniques address the repairs, mitigations, crack growth, and various damage conditions in polymer matrix. These techniques have advantages and limitations specific to applications that are summarized as follows:

(i) These have slow healing rate.

(ii) Additional requirement of heat/light in intrinsic systems can be good or bad depending on the application.

(iii) The use of foreign inserts in matrix leads to detrimental effects on matrix.

(iv) Complexity of vascular networks is a challenge.

(v) Do not address the ballistic or hypervelocity impacts, which are great promising to space exploration upon material puncture healing.

4.2 Supramolecular noncovalent interaction-based self-healing

In materials, microscopic damages are repaired by extrinsic technology in which foreign species play a lead role in the healing process. These techniques respond to damages autonomically or stimulus-assisted phenomenon and take a shorter time to recover the strength of materials. But in the case of macroscopic damages, these extrinsic techniques are poorer in performance. Additionally, structurally covalent dynamic polymers are also requested with additional heat to clear microdamages. To overcome these issues, some significant research in the field of supramolecular systems is focused which respond to damages autonomously and recovered mechanical integrity without the addition of foreign reactive species and human intervention. In high-performance materials, macroscopic and energetic damage events are usual. These damages are healed by physical interactions. The physical interactions are noncovalent in nature and cause autonomic healing due to inherent origin. These recover about fully mechanical properties but take longer time. These noncovalent interactions recover mechanical properties almost completely but take longer time. These interactions are reversible subjected to the thermodynamic equilibrium and show additional impacts such as environmental-dependent switch properties, easy processability

S.N.	Nature of interactions	Healing system	Ref.	Remarks
1	Hydrogen bonding	DCPD/DNE/epoxy systems	[22]	Adhesion promoter used
		Polystyrene grafted with poly(acrylate amide)	[23]	Polyvalent H-bonding sites
		Ureidopyrimidinone bond—SupraPolix BV	[24]	Self-assembly mechanism
2	π-π interaction	Polydiimide/poly(siloxane) with pyrenyl end groups	[25, 26]	Flexible and self-supporting material
		Copolyimide/poly(amide) with pyrenyl end groups	[66]	Thermoreversible system
3	Ionomers	Poly(ethylene-co-methacrylic acid)/Na + ion (EMAA)	[27–30]	Self-sealing shooting range targets, tires

Table 3.
Typical supramolecular interactions and relevant self-healing polymer systems.

and self-healing behavior as compared to traditional polymers. Reversible bonding can be used to design supramolecular healable polymers and composites which respond to external stimuli such as heat [62], pressure [63], water [64], or light [65]. Various supramolecular interaction-based self-healing systems are shown in **Table 3**.

Stimuli responsiveness [67] and a high diffusion rate of oligomeric components [68] are the main characteristic of supramolecular polymers which make them for rapid and controllable healing system. Oligomers are low-molecular-weight species that make aggregates by self-assembling and perform rheological or mechanical properties similar as polymers. The reversible non-covalent interactions can be possible by hydrogen bonding, ionic bonding, metal-ligand bonding, and π-π stacking.

4.2.1 Supramolecular chemistry based on H-bonding

H-bonding is the most popular route to achieve supramolecular polymers. Upon heat, interactions between the polymeric chains are decreased and reassembled upon cooling, and finally the non-covalent cross-linking recovered the strength and mechanical integrity. To achieve sufficient cross-linking density in polymer, a high association constant between repeating units is needed. The association constant and a reversible interaction have a reverse relation. In contrast, at the less association constant, better reversibility is achieved but having smaller assemblies and poor mechanical properties. However, individual supramolecular polymers are suffered by low mechanical strength. The mechanical strength of supramolecular system is enhanced by an increase in the number of non-covalent interaction [22] and by reinforcement with nanofiller [26]. Significantly, the interactions of matrix fiber are increased by the presence of hydrogen bond accepting functionality in matrix polymer blend. Adhesion promoter increases the amount of H-bonding of the matrix system [22]. These promoters are known as co-healing agents shown in **Figure 8**. On fracture of capsules, the DCPD monomer penetrates the networks of epoxy matrix and reacts with embedded catalyst. The formation of interpenetrated network is initiated by polymerization of DCPD which would strengthen the matrix—poly(DCPD) surface. Higher strength is obtained by entangled

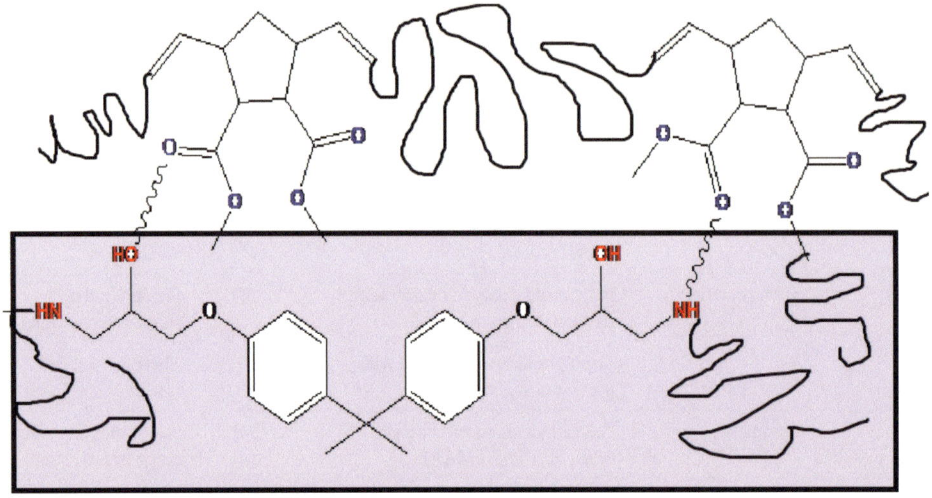

Figure 8.
Supramolecular interactions with adhesion promoter [22].

networks, and H-bonding will also improve overall bonding strength and healing efficiency. H-bonding-incorporated supramolecular system is shown in **Figure 8**.

The reinforcement of cellulose nanocrystals or cellulose nanowhiskers into polymer matrix such as poly(ethylene oxide-co-epichlorohydrin) [69] and low-density polyethylene [70] has shown improved stiffness corresponding to parental matrix materials.

4.2.2 Metal-ligand supramolecular polymers

Optical and photo physical properties of metal complexes offer to design advanced materials. Reversible behavior of metal ion and ligand bond in metal-ligand complexes attract the research community to design stimuli-responsive materials. Metallo-supramolecular polymers have low-molecular-weight species known as telechelic. These are attached with ligand end group through metal-ion linkage. These polymers can be healed upon contact of light [65, 71]. During the whole healing mechanism, supramolecular interactions and light-heat conversion happen subsequently. The temporary disentanglement of metal-ligand motifs is possible when excited electronically upon contact with UV, and further, heat energy is released. Subsequently, the average molecular weight and viscosity of system are decreased, and defect healing is resulted. Local damages can also be recovered just upon light exposure. A metallo-hydrogel based on histidine and Zn^{2+} ions is designed using coordination-driven self-assembly [72]. The hydrogel formation is instantaneous, and it exhibits stimuli-responsive behavior with respect to pH, heat, and external chemicals.

4.2.3 Supramolecular π-π interaction-assisted self-healing

A thermally triggered reversible network is achieved based on π-π stacking interactions in which end-capped π-electron-deficient groups interact with π-electron-rich aromatic backbone. The chain-folding co-polyimide (electron deficient) and pyrenyl (electron rich) end-capped polyamide chains have such π-π interactions [66]. The driving forces for producing tough, stable, healable homogeneous blend of elastomers are interpolymeric π-π stacking complexes [73]. At higher temperatures, the disengagement of the supramolecular (π-π stacking and hydrogen-bonding) interactions is possible, which leads to change in the apparent molecular

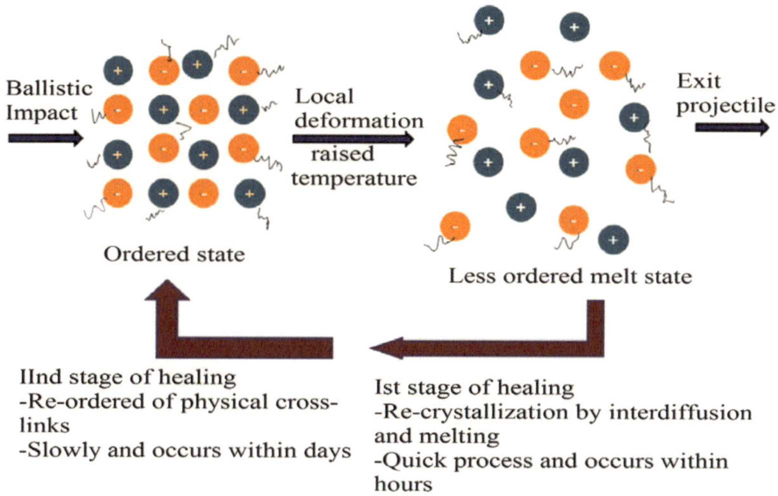

Figure 9.
Schematic diagram of ionomeric healing upon ballistic impact based on order-disorder theory of healing presented by Fall [27].

weight of the homogeneous noncovalent polymer blend and further rapid change in viscosity with temperature. A polymer blend recovered the mechanical strength by refurbishing these supramolecular interactions.

4.2.4 Supramolecular self-healing ionomers

Ionomers are polymers in which the bulk properties are governed by ionic interactions in discrete regions of the material [74]. Ionomers contain up to 15% ionic groups and respond instantaneously and autonomously in absence of external species, and additional heat or other stimuli make ionomers unique. Various bonding interactions such as ionic, dipole-dipole, or ion-dipole bonds are key factors to develop self-healing systems based on ionomers. Ballistic healing proceeds through combination of an elastic response (i.e., attain pure shape) and a viscous response (secondary polymer flow and chain entanglement) of intrinsic aggregates. Initially after impact, projectile transfers some of the impact energy to ionomer system which melts the matrix, and rest kinetic energy is stored elastically to movement, and the projectile is ejected and leaves behind the matrix with some melted portion. Finally, the hole is sealed and recovered some mechanical properties followed by crystallization and reaggregation of ionomers at the damage site. Healing process in ionomers is multistep. Initially upon high-energy impacts, the local deformation proceeded to less ordered melt state and resultant the projectile is ejected. After that, complex aggregates are formed via recrystallization of interdiffusion of intermolecular interactions. It is a quick process that happens within seconds to hours. In the second step, restructuring of physical cross-linking leads the final stage of healing; it takes longer duration as usually days to months. Some of commercial products such as React-A-Seal, Surlyn, and Nucrel have EMAA as base matrix [27]. It is a copolymer of ethylene and a vinyl monomer with an acidic group. EMAA ionomer exposed self-healing upon ballistic impact (**Figure 9**) [27, 28].

During whole mechanism "free volume" plays a great role which provides enough mobility to polymer chain rearrangement and interdiffusion. Besides, many other factors, such as impact energy, nature of ionic groups, and counterions, the neutralization degree, increased temperature during impact, the content of ionic groups and dielectric constant, and so on, also play a key role to succeed self-healing by ionomers. To enhance the healing efficiency of ionomers composites, conductive

S.N.	Healing system	Healing mechanism (photoinduced)	Ref.	Remarks
1	1,1,1-tris(cinnamoyloxy-methyl) ethane (TCE) monomer	[2 + 2] Cycloaddition	[40]	Cycloaddition of cinnamoyl groups
2	Pentaerythritol tetra (3-mercaptopropionate) and triethylene glycol divinyl ether system	Reversible cleavage of allyl sulfide linkage	[76]	Hemolytic photolysis
3	Pyridine-capped poly(methyl acrylate)s/SCS-cyclometalated dipalladium	Light-induced cis-trans isomerization	[77]	Molecular switching
4	Epoxy/SiO2 microcapsule	UV-triggered polymerization	[78]	Aerospace coatings
5	Chitosan polyurethane networks	Photoinduced reaction	[79]	Coating applications
6	Epoxy/nano-TiO2	Photoabsorbing microcapsules	[80]	Protection coatings

Table 4.
Photoinduced damage-healing system.

and magnetic fillers are added [29]. In addition to plasticizer such as zinc stearate ionic domain, the matrix properties are intact [75].

5. Photochemically induced healing chemistry

Photoinduced healing is currently demanding because of rapid, eco-friendly concept of healing of polymer matrix composites. It is induced by the application of a strong light irradiation. To damage healing, the foreign healing inserts, catalyst, or additional heat is neither required to comply with the healing process. Such system based on photochemical reactions is developed (**Table 4**).

6. Significance of healing functionality

The objective of self-healing phenomenon is focused to respond upon fracture based on quasi-static performance. In the future, self-healing materials can be used in various applications including nuclear reactors, marine structures, and various cutting tools. Self-healing aircrafts or spacecraft is currently desired in references of micrometeoroids and orbital debris protection, low-temperature repairs for aero-structures, and re-healable fuel tanks. Cost-efficient healable wind power blades and wire insulation materials are also desirable based upon self-healing materials. Durable targets for military personnels and training are also part of self-healing ionomers which protect against ballistic penetrations.

7. Current scenario, challenges, and future perspective

Self-healing systems represent a wide interdisciplinary area and are performed by different techniques. Understanding the basic mechanism and principle of damage healing is also still a challenge for science community. Healing chemistries have greater stabilization, higher reactivity, and fast kinetics, although many challenges still remain. Healing performance under long term in drastic environment is advantageous. Additionally, increased healing efficiency with low-cost techniques and least detrimental effect on base matrix is also challenging. To implement encapsulation technique, distribution of capsules is challenging and has adverse effects on matrix. Similarly, the fabrication of vascular network is not easy to implement. In microelectronics, optoelectronics, and robotics where damages are common as nanostructures, self-healing concept is beneficiary. Long-lasting self-healing battery, self-sealing fuel tank/tires, fad-resistant textiles, and anti-tamper electronics are future significance of self-healing technique. Sensor design with healing concept is used to structural health monitoring of components. The incorporation of self-healing concept into aerostructures is challenging and required to establish because the aerostructures always suffered mechanical and thermal loading continuously. In the future, catalyst-free and UV- or light-oriented self-healing systems are desirable.

8. Conclusive remarks

Conclusively, the self-healing techniques explore the possibility of damage detection and subsequent damage management in materials. Some technological restrictions and possible solutions are also explored. Self-healing techniques intimate damages at the early stage and recovered mechanical properties at some extent which reduces the probability of replacement and loss of structures. Currently, we are still waiting for fully self-healing systems that can show both stimuli responsiveness and healing characteristics.

Acknowledgements

The author would like to say thanks to Dr. Dibeyndu S. Bag and Dr. Ajit Shankar Singh, DMSRDE, Kanpur, for their adequate suggestion to design the article.

Author details

Jeetendra Kumar Banshiwal* and Durgesh Nath Tripathi
Defence Materials and Stores Research and Development Establishment, Kanpur, India

*Address all correspondence to: jeetukumar331@gmail.com

References

[1] Mercier P. Material for Protecting Vessels, Receptacles. & c US Patent Specification. 1896. p. 561

[2] Schwartz SS. Self-Sealing Space Suit. U.S. Patent; 3,536576. 1970

[3] White SR et al. Autonomic healing of polymer composites. Nature. 2001;**409**(6822):794. DOI: 10.1038/35057232

[4] Jin H et al. Self-healing thermoset using encapsulated epoxy-amine healing chemistry. Polymer. 2012;**53**(2):581-587. DOI: 10.1016/j.polymer.2011.12.005

[5] Kamphaus JM et al. A new self-healing epoxy with tungsten (VI) chloride catalyst. Journal of the Royal Society Interface. 2007;**5**(18):95-103. DOI: 10.1098/rsif.2007.1071

[6] Yuan YC et al. Self-healing polymeric materials using epoxy/mercaptan as the healant. Macromolecules. 2008;**41**(14):5197-5202. DOI: 10.1021/ma800028d

[7] Cho SH et al. Polydimethylsiloxane-based self-healing materials. Advanced Materials. 2006;**18**(8):997-1000. DOI: 10.1002/adma.200501814

[8] Yin T et al. Durability of self-healing woven glass fabric/epoxy composites. Smart Materials and Structures. 2009;**18**(7):074001. DOI: 10.1088/0964-1726/18/7/074001

[9] Bond IP, Williams MGJ, Trask RS. Self-healing CFRP for Aerospace Applications. 16th ed. Advanced Composites Centre for Innovation and Science, University of Bristol, Department of Aerospace Engineering, Bristol, UK. 2007

[10] Wu XF et al. Electrospinning core-shell nanofibers for interfacial toughening and self-healing of carbon-fiber/epoxy composites. Journal of Applied Polymer Science. 2013;**129**(3):1383-1393. DOI: 10.1002/app.38838

[11] Bleay SM et al. A smart repair system for polymer matrix composites. Composites Part A: Applied Science and Manufacturing. 2001;**32**(12):1767-1776. DOI: 10.1016/S1359-835X(01)00020-3

[12] Toohey KS et al. Delivery of two-part self-healing chemistry via microvascular networks. Advanced Functional Materials. 2009;**19**(9): 1399-1405. DOI: 10.1002/adfm.200801824

[13] Williams HR, Trask RS, Bond IP. Self-healing composite sandwich structures. Smart Materials and Structures. 2007;**16**(4):1198. DOI: 10.1088/0964-1726/16/4/031

[14] Kousourakis A, Mouritz A. The effect of self-healing hollow fibres on the mechanical properties of polymer composites. Smart Materials and Structures. 2010;**19**(8):085021. DOI: 10.1088/0964-1726/19/8/085021

[15] Toohey KS et al. Self-healing materials with microvascular networks. Nature Materials. 2007;**6**(8):581. DOI: 10.1038/nmat1934

[16] Wang Y et al. Sustainable self-healing at ultra-low temperatures in structural composites incorporating hollow vessels and heating elements. Royal Society Open Science. 2016;**3**(9):160488. DOI: 10.1098/rsos.160488

[17] Murphy EB et al. Synthesis and characterization of a single-component thermally remendable polymer network: Staudinger and Stille revisited. Macromolecules. 2008;**41**(14): 5203-5209. DOI: 10.1021/ma800432g

[18] Chen X et al. A thermally re-mendable cross-linked polymeric material. Science. 2002;**295**(5560):

1698-1702. DOI: 10.1126/science.1065879

[19] Dumitrescu S, Grigoras M, Natansohn A. Diels-Alder polymerization of two anthracene monomers. Journal of Polymer Science, Polymer Letters Edition. 1979;**17**(9): 553-559. DOI: 10.1002/pol.1979.130170902

[20] Chen X et al. New thermally remendable highly cross-linked polymeric materials. Macromolecules. 2003;**36**(6):1802-1807. DOI: 10.1021/ma0210675

[21] Kolb HC, Finn M, Sharpless KB. Click chemistry: Diverse chemical function from a few good reactions. Angewandte Chemie International Edition. 2001;**40**(11):2004-2021. DOI: 10.1002/1521-3773 (20010601) 40:11<2004::AID-ANIE2004>3.0.CO;2-5

[22] Wilson GO et al. Adhesion promotion via noncovalent interactions in self-healing polymers. ACS Applied Materials & Interfaces. 2011;**3**(8): 3072-3077. DOI: 10.1021/am200584z

[23] Chen Y et al. Multiphase design of autonomic self-healing thermoplastic elastomers. Nature Chemistry. 2012;**4**(6):467-472. DOI: 10.1038/nchem.1314

[24] Bosman AW, Sijbesma RP, Meijer E. Supramolecular polymers at work. Materials Today. 2004;**7**(4):34-39. DOI: 10.1016/S1369-7021(04) 00187-7

[25] Burattini S et al. A supramolecular polymer based on tweezer-type π–π stacking interactions: Molecular design for healability and enhanced toughness. Chemistry of Materials. 2010;**23**(1):6-8. DOI: 10.1021/cm102963k

[26] Fox J et al. High-strength, healable, supramolecular polymer nanocomposites. Journal of the American Chemical Society. 2012;**134**(11):5362-5368. DOI: 10.1021/ja300050x

[27] Fall RA. Puncture Reversal of Polyethylene Ionomers-Mechanistic Studies. The faculty of Virginia Polytechnic Institute and State University Blacksburg: Virginia Tech; 2001

[28] Kalista SJ, Ward TC. Thermal characteristics of the self-healing response in poly (ethylene-co-methacrylic acid) copolymers. Journal of the Royal Society Interface. 2007;**4**(13):405-411. DOI: 10.1098/rsif.2006.0169

[29] Owen CC. Magnetic Induction for in-situ Healing of Polymeric Material. Virginia Tech. 2006

[30] Seibert GM. Shooting Range Target. Google Patents. U.S. Patent No. 5,486,425. 23 Jan. 1996

[31] Yang Y, Urban MW. Self-healing polymeric materials. Chemical Society Reviews. 2013;**42**(17):7446-7467. DOI: 10.1039/C3CS60109A

[32] Flory P, Krigbaum W. Statistical mechanics of dilute polymer solutions. II. The Journal of Chemical Physics. 1950;**18**(8):1086-1094. DOI: 10.1063/1.1747866

[33] Liu F, Urban MW. Recent advances and challenges in designing stimuli-responsive polymers. Progress in Polymer Science. 2010;**35**(1-2):3-23. DOI: 10.1016/j.progpolymsci.2009.10.002

[34] Park JS et al. Multiple healing effect of thermally activated self-healing composites based on Diels–Alder reaction. Composites Science and Technology. 2010;**70**(15):2154-2159. DOI: 10.1016/ j.compscitech. 2010. 08. 017

[35] Nji J, Li G. A self-healing 3D woven fabric reinforced shape memory polymer composite for impact mitigation. Smart Materials and

Structures. 2010;**19**(3):035007. DOI: 10.1088/0964-1726/19/3/035007

[36] Hsieh H-C, Yang T-J, Lee S. Crack healing in poly (methyl methacrylate) induced by co-solvent of methanol and ethanol. Polymer. 2001;**42**(3):1227-1241. DOI: 10.1016/S0032-3861(00)00407-9

[37] Wool R, O'connor K. A theory crack healing in polymers. Journal of Applied Physics. 1981;**52**(10):5953-5963. DOI: 10.1063/1.328526

[38] Wu DY, Meure S, Solomon D. Self-healing polymeric materials: A review of recent developments. Progress in Polymer Science. 2008;**33**(5):479-522. DOI: 10.1016/j.progpolymsci.2008.02.001

[39] Kryger MJ et al. Masked cyanoacrylates unveiled by mechanical force. Journal of the American Chemical Society. 2010;**132**(13):4558-4559. DOI: 10.1021/ja1008932

[40] Chung C-M et al. Crack healing in polymeric materials via photochemical [2 + 2] cycloaddition. Chemistry of Materials. 2004;**16**(21):3982-3984. DOI: 10.1021/cm049394+

[41] Davis DA et al. Force-induced activation of covalent bonds in mechanoresponsive polymeric materials. Nature. 2009;**459**(7243):68. DOI: 10.1038/nature07970

[42] Karthikeyan S, Sijbesma RP. Mechanochemistry: Forcing a molecule's hand. Nature Chemistry. 2010;**2**(6):436. DOI: 10.1038/nchem.677

[43] Blaiszik B, Sottos N, White S. Nanocapsules for self-healing materials. Composites Science and Technology. 2008;**68**(3):978-986. DOI: 10.1016/j.compscitech.2007.07.021

[44] Caruso MM et al. Robust, double-walled microcapsules for self-healing polymeric materials. ACS Applied Materials & Interfaces. 2010;**2**(4): 1195-1199. DOI: 10.1021/am100084k

[45] Keller MW, White SR, Sottos NR. A self-healing poly (dimethyl siloxane) elastomer. Advanced Functional Materials. 2007;**17**(14):2399-2404. DOI: 10.1002/adfm.200700086

[46] Billiet S et al. Development of optimized autonomous self-healing systems for epoxy materials based on maleimide chemistry. Polymer. 2012;**53**(12):2320-2326. DOI: 10.1016/j.polymer.2012.03.061

[47] Rule JD et al. Wax-protected catalyst microspheres for efficient self-healing materials. Advanced Materials. 2005;**17**(2):205-208. DOI: 10.1002/adma.200400607

[48] Chung US et al. Polyurethane matrix incorporating PDMS-based self-healing microcapsules with enhanced mechanical and thermal stability. Colloids and Surfaces A: Physicochemical and Engineering Aspects. 2017;**518**:173-180. DOI: 10.1016/j.colsurfa.2017.01.044

[49] Khun NW et al. Wear resistant epoxy composites with diisocyanate-based self-healing functionality. Wear. 2014;**313**(1):19-28. DOI: 10.1016/j.wear.2014.02.011

[50] Li Q et al. Effects of processing conditions of poly (methylmeth-acrylate) encapsulated liquid curing agent on the properties of self-healing composites. Composites Part B: Engineering. 2013;**49**:6-15. DOI: 10.1016/j.compositesb.2013.01.011

[51] Khan NI, Halder S, Goyat MS. Influence of dual-component microcapsules on self-healing efficiency and performance of metal-epoxy composite-lap joints. The Journal of Adhesion. 2016:**93**(12):949-963. DOI: 10.1080/00218464.2016.1193806

[52] Dry C, Sottos N. Passive smart self-repair in polymer matrix composite materials. In: Conference of adaptive materials, Albuquerque, New Mexico. January 1993. SPIE Vol. 1916, 438-444. DOI: 10.1117/12.148501

[53] Fifo O, Ryan K, Basu B. Glass fibre polyester composite with in vivo vascular channel for use in self-healing. Smart Materials and Structures. 2014;**23**(9):095017. DOI: 10.1088/0964-1726/23/9/095017

[54] Pang JW, Bond IP. A hollow fibre reinforced polymer composite encompassing self-healing and enhanced damage visibility. Composites Science and Technology. 2005;**65**(11):1791-1799. DOI: 10.1016/j.compscitech.2005.03.008

[55] Garcia S, Fischer H, Román J. Self-healing polymer systems: Properties, synthesis and applications. Smart Polymers and their Applications. 2014:271-298. DOI: 10.1533/9780857097026.1.271

[56] An SY et al. Dual sulfide–disulfide crosslinked networks with rapid and room temperature self-healability. Macromolecular Rapid Communications. 2015;**36**(13):1255-1260. DOI: 10.1002/marc.201500123

[57] Fang Z et al. Healable, Reconfigurable, Reprocessable Thermoset Shape Memory Polymer with Highly Tunable Topological Rearrangement Kinetics. ACS Applied Materials & Interfaces. 2017;**9**(27):22077-22082. DOI: 10.1021/acsami.7b05713

[58] Telitel S et al. Introduction of self-healing properties into covalent polymer networks via the photodissociation of alkoxyamine junctions. Polymer Chemistry. 2014;**5**(3):921-930. DOI: 10.1039/C3PY01162C

[59] Wu S et al. Ultrafast self-healing nanocomposites via infrared laser and their application in flexible electronics. ACS Applied Materials & Interfaces. 2017;**9**(3):3040-3049. DOI: 10.1021/acsami.6b15476

[60] Chujo Y et al. Synthesis and redox gelation of disulfide-modified polyoxazoline. Macromolecules. 1993;**26**(5):883-887. DOI: 10.1021/ma00057a001

[61] Tsarevsky NV, Matyjaszewski K. Reversible redox cleavage/coupling of polystyrene with disulfide or thiol groups prepared by atom transfer radical polymerization. Macromolecules. 2002;**35**(24):9009-9014. DOI: 10.1021/ma021061f

[62] Cordier P et al. Self-healing and thermoreversible rubber from supramolecular assembly. Nature. 2008;**451**(7181):977. DOI: 10.1038/nature06669

[63] Haraguchi K, Uyama K, Tanimoto H. Self-healing in nanocomposite hydrogels. Macromolecular Rapid Communications. 2011;**32**(16):1253-1258. DOI: 10.1002/marc.201100248

[64] South AB, Lyon LA. Autonomic self-healing of hydrogel thin films. Angewandte Chemie International Edition. 2010;**49**(4):767-771. DOI: 10.1002/anie.200906040

[65] Burnworth M et al. Optically healable supramolecular polymers. Nature. 2011;**472**(7343):334. DOI: 10.1038/nature09963

[66] Burattini S et al. A self-repairing, supramolecular polymer system: Healability as a consequence of donor–acceptor π–π stacking interactions. Chemical Communications. 2009;**44**:6717-6719. DOI: 10.1039/b910648k

[67] Fox JD, Rowan SJ. Supramolecular polymerizations and

main-chain supramolecular polymers. Macromolecules. 2009;**42**(18): 6823-6835. DOI: 10.1021/ma901144t

[68] Kim YH, Wool RP. A theory of healing at a polymer-polymer interface. Macromolecules. 1983;**16**(7):1115-1120. DOI: 10.1021/ma00241a013

[69] Capadona JR et al. Stimuli-responsive polymer nanocomposites inspired by the sea cucumber dermis. Science. 2008;**319**(5868):1370-1374. DOI: 10.1126/science.1153307

[70] Ben Azouz K et al. Simple method for the melt extrusion of a cellulose nanocrystal reinforced hydrophobic polymer. ACS Macro Letters. 2011;**1**(1):236-240. DOI: 10.1021/mz2001737

[71] Biyani MV, Foster EJ, Weder C. Light-healable supramolecular nanocomposites based on modified cellulose nanocrystals. ACS Macro Letters. 2013;**2**(3):236-240. DOI: 10.1021/mz400059w

[72] Sharma B et al. Chirality control of multi-stimuli responsive and self-healing supramolecular metallo-hydrogels. New Journal of Chemistry. 2018;**42**(8):6427-6432. DOI: 10.1039/C8NJ00218E

[73] Burattini S et al. A healable supramolecular polymer blend based on aromatic π–π stacking and hydrogen-bonding interactions. Journal of the American Chemical Society. 2010;**132**(34):12051-12058. DOI: 10.1021/ja104446r

[74] Eisenberg A, Rinaudo M. Polyelectrolytes and ionomers. Polymer Bulletin. 1990;**24**(6):671-671

[75] Varley RJ, van der Zwaag S. Development of a quasi-static test method to investigate the origin of self-healing in ionomers under ballistic conditions. Polymer Testing. 2008;**27**(1):11-19. DOI: 10.1016/j.polymertesting.2007.07.013

[76] Scott TF et al. Photoinduced plasticity in cross-linked polymers. Science. 2005;**308**(5728):1615-1617. DOI: 10.1126/science.1110505

[77] Tennyson AG, Wiggins KM, Bielawski CW. Mechanical activation of catalysts for C–C bond forming and anionic polymerization reactions from a single macromolecular reagent. Journal of the American Chemical Society. 2010;**132**(46):16631-16636. DOI: 10.1021/ja107620y

[78] Guo W et al. UV-triggered self-healing of a single robust SiO_2 microcapsule based on cationic polymerization for potential application in aerospace coatings. ACS Applied Materials & Interfaces. 2016;**8**(32):21046-21054. DOI: 10.1021/acsami.6b06091

[79] Ghosh B, Urban MW. Self-repairing oxetane-substituted chitosan polyurethane networks. Science. 2009;**323**(5920):1458-1460. DOI: 10.1126/science.1167391

[80] Gao L et al. Photoresponsive self-healing polymer composite with photoabsorbing hybrid microcapsules. ACS Applied Materials & Interfaces. 2015;**7**(45):25546-25552. DOI: 10.1021/acsami.5b09121

Chapter 3

Janus Nanosheets Derived from $K_4Nb_6O_{17}·3H_2O$ *via* Regioselective Interlayer Surface Modification

Ryoko Suzuki, Mitsuhito Sudo, Megumi Hirano, Naokazu Idota, Masashi Kunitake, Taisei Nishimi and Yoshiyuki Sugahara

Abstract

Inorganic Janus nanosheets were successfully prepared using the difference in reactivity between interlayers I and II of layered hexaniobate $K_4Nb6O_{17}·3H_2O$. Janus nanosheets exhibit the highest anisotropy among Janus compounds due to their morphology. It is therefore important to prepare Janus nanosheets with stable shapes in various solvents, robust chemical bonds between nanosheets and functional groups and high versatility due to surface functional groups. $K_4Nb_6O_{17}·3H_2O$, which possesses two types of interlayers and two types of organophosphonic acids that react with metal oxides to form robust covalent bonds, was employed to prepare Janus nanosheets for this study. Interlayer I was modified by octadecylphosphonic acid, followed by modification by carboxypropylphosphonic acid mainly at interlayer II. Preparation of Janus nanosheets with two organophosphonate moieties was confirmed by ^{31}P MAS NMR. After these regioselective and sequential modifications, the products were exfoliated into single-layered nanosheets in THF. Two types of derivatives with different repeating distances were recovered from a dispersion containing nanosheets exfoliated by different processes, centrifugation, and solvent evaporation. AFM analysis of the exfoliated nanosheets revealed that the products were Janus compounds. There are high expectations for application of these types of Janus nanosheets in various fields and for design of various Janus nanosheets using this preparation method.

Keywords: Janus nanosheets, $K_4Nb_6O_{17}·3H_2O$, organophosphonic acid, grafting reaction, intercalation

1. Introduction

A Janus compound has two surface properties, and each of these properties appears on one of two sides of the compound [1]. Janus compounds are expected to be applied as functional materials, including electronic paper [2], solid surfactants [3], optics materials [4], and drug delivery system (DDS) vectors [5]. The morphology of Janus compounds is classified into three categories: 0-dimensional compounds such as particles; one-dimensional compounds including cylinders, tubes, and rods; and two-dimensional compounds, typically sheets or discs [6–10].

There are various methods of preparing Janus nanoparticles. Regioselective surface modification of nanoparticles can produce Janus nanoparticles [11]. By forming nanoparticles with two raw materials, Janus nanoparticles with different compositions can be prepared [2]. Self-assembly and subsequent cross-linking of polymer chains can also produce Janus nanoparticles [12]. Janus rods can be prepared by forming silica using TEOS by the sol-gel method on Fe_3O_4 regioselectively [13]. Janus cylinders can be prepared by masking one side of a cylinder and conducting subsequent modification of the other side [14].

Janus nanosheets exhibit the highest anisotropy among Janus compounds due to their unique morphology. They are also useful as emulsifiers, because Janus nanosheets cannot rotate at the interfaces of micelles [15]. Most Janus nanosheets reported so far have consisted of polymers. Stupp et al. reported the preparation of Janus nanosheets by polymerization of oligomers with polymerizable groups [16]. Polymerization of oligomers led to the formation of sheet morphology, because the polymerizable groups were located at the center of an oligomer. Walther et al. used triblock copolymers, polystyrene-*block*-polybutadiene-*block*-poly(*tert*-butyl methacrylate), for preparing Janus nanosheets [17]. Janus nanosheets were prepared by cross-linking polybutadiene domains of the triblock copolymers, and the resulting Janus nanosheets had two types of surfaces, polystyrene, and poly(*tert*-butyl methacrylate) moieties. These methods are based on selective polymerization or cross-linking. On the other hand, Janus nanosheets were prepared by dropping poly(ε-caprolactone) at the interface between water and pentyl acetate and evaporating solvents to crystallize polymers [10]. The properties of Janus compounds were realized by folding the polymer an odd number of times, which exposed carboxyl groups on one side of the nanosheets. Although polymer-based Janus nanosheets have been prepared, these Janus nanosheets became swollen and deformed in organic solvents because they were composed of polymers [18].

It is therefore obvious that the preparation of inorganic Janus nanosheets is an important issue. There are only a few reports of preparation of Janus nanosheets derived from inorganic compounds, however, comprising silica nanosheets prepared by the sol-gel method [19–21] and fluorohectorite, which is a layered clay mineral [22].

Silica-based Janus nanosheets were prepared by hydrolysis and condensation of two trifunctional organoalkoxysilanes. First, phenylalkoxysilane was assembled on the surface of oil in a W/O emulsion and reacted by the sol-gel method. The product, which formed a hollow shell, was further reacted with aminoalkyltrialkoxysilane. The resultant product had phenyl groups and amino groups on the inner and outer surfaces of hollow particles, respectively. Janus nanosheets were obtained by breaking the hollow silica particles using a colloid milling method. The resulting Janus nanosheets have a thickness of 65 nm and a curvature originating from the hollow particle morphology [19]. Another method of preparing silica-based Janus nanosheets using a $CaCO_3$ template was also reported. First, 3-butyldianhydride mercaptopropyltrimethoxysilane was assembled on the surface of $CaCO_3$ particles and reacted using a sol-gel process. The products were further reacted with octadecyltrichlorosilane. Janus nanosheets were obtained by removing the templates and crushing the resultant hollow particles. Janus nanosheets with single a nanometer thickness were prepared by this method [21]. These sol-gel preparation methods, which were developed by Yang et al., required adsorption of organosilanes on liquid-liquid interfaces or self-assembly on templates. Thus, these methods restricted the reaction system design and choices of molecules.

Another method of preparing Janus nanosheets using inorganic layered material was reported in which the Janus nanosheets consisted of a sheet of a clay mineral, fluorohectorite, and two cationic polymers. The interlayer of fluorohectorite was selectively exfoliated into double-layered nanosheets. Both surfaces of the products, which had a negative charge, were modified with a cationic polymer, protonated polyethyleneimine-ethylene oxide. After that, the products were exfoliated into single-layered nanosheets, and one surface which had not been modified was reacted with another cationic polymer, protonated dendritic poly(amidoamine). The products had one cationic polymer on one side and the other polymer on the other side of the nanosheets [22]. This preparation method has the advantage that any cationic compound could be used as the modifier, regardless of its functional group. On the other hand, the products use in water was restricted due to the presence of ionic bonds between the surfaces of the clay sheets and the polymers.

Preparation of graphene-based Janus nanosheets has also been developed [23]. For example, two-step functionalization was carried in one report [24]. First, the graphene surface of the substrate was functionalized by a photochlorination reaction. The functionalized graphene was then peeled off the substrate to expose the unmodified side using PMMA film as a mediator. The exposed fresh side of the graphene was further functionalized by phenylation reaction. Graphene-based Janus nanosheets were also prepared using a Pickering emulsion [25]. A graphene oxide (GO) dispersion was mixed with hydrochloric acid and wax, and this mixture was ultrasonicated to prepare a Pickering emulsion. The micelles were then washed with a sodium hydroxide aqueous solution, and GO was adsorbed on the surface of micelles to form a monolayer. The exposed GO surface of the micelles was further modified with alkylamine. Finally, Janus nanosheets were obtained by dissolving wax in chloroform. In the preparation of graphene-based Janus nanosheets, it is necessary that a single layer of graphene be adsorbed on a substrate or at the liquid-liquid interphase to achieve regioselective functionalization.

These earlier studies show that the following conditions are desirable: nanosheets should have covalent bonds with organic groups, the choice of functional groups should not be limited, and regioselective surface modification should be easily achieved. Another preparation method that satisfies the above conditions should therefore be developed.

Some inorganic layered materials have structures in which negatively charged nanosheets and metal cations are piled up alternately. Nanosheets are very thin, just a few atoms thick. On the other hand, their lateral sizes are large, possibly on a micrometer scale. Therefore, these materials have high aspect ratios. From the viewpoints of structure and composition, inorganic layered materials are classified into several categories, including clay minerals [26–28], layered silicate [29–31], and layered transition metal oxide [32–34]. These inorganic layered materials have been used as hosts for inorganic-organic hybrids. There are only two types of reaction for preparing inorganic-organic hybrids: intercalation reactions forming noncovalent bonds and grafting reactions forming covalent bonds. Grafting reactions that form covalent bonds between layer surfaces and organic groups utilize alcohols, carboxylic acids, silane coupling agents, and phosphorous coupling reagents such as organophosphonic acids as modifiers. Grafted organophosphonate moieties, in particular, are seldom eliminated from nanosheet surfaces because organophosphonate moieties form stable M–O–P bonds with nanosheet surfaces, except for Si–O–P bonds, which are known to become hydrolyzed under certain conditions [35]. Surface modification with monolayer can be easily achieved, moreover because homocondensation reactions do not occur between organophosphonic acids under mild conditions [36]. There have been many reports of surface modification by silane coupling agents for clay minerals [37], layered silicates [38–40], and layered transition metal oxides [41, 42]. In the case of surface modification by alcohol, there have been reports of polysilicates with ethylene glycol [43] and aliphatic alcohols [44], while layered perovskites have been modified with n-alcohol [45] and alcohol with fluoroalkyl groups [44, 46]. Also, layered perovskites were grafted with phenyl or *n*-alkylphosphonic acids using the aforementioned *n*-alkoxy derivatives as intermediates [47].

Layered hexaniobate ($K_4Nb_6O_{17}{\cdot}3H_2O$) has a unique structure among layered transition metal oxides; $K_4Nb_6O_{17}{\cdot}3H_2O$ has two types of interlayers that are piled up alternately and exhibit different reactivities [48]. Interlayer I possesses hydrated water and shows high reactivity, which anhydrous interlayer II exhibits low reactivity. There have been a certain number of reports of reactions between $K_4Nb_6O_{17}{\cdot}3H_2O$ and organic molecules using the differences in reactivity between interlayer I and interlayer II [33].

Intercalation of small ammonium ions occurred sequentially, first in interlayer I and then in interlayer II [49]. On the other hand, bulky ammonium ions were intercalated only into interlayer I [50]. Compounds that were modified only in interlayer I were called A-type, and compounds that were modified in both interlayers I and II were called B-type. Kimura et al. modified the $K_4Nb_6O_{17}{\cdot}3H_2O$ surfaces with phenylphosphonic acid using A-type and B-type ion-exchanged intercalation compounds of $K_4Nb_6O_{17}{\cdot}3H_2O$ [51]. In their report, bulky dioctadecyldimethylammonium ions were intercalated into only interlayer I, and A-type phenylphosphonate derivatives were obtained using this A-type ammonium intercalation compound as an intermediate. On the other hand, dodecylammonium ions were intercalated into both interlayers I and II and a B-type phenylphosphonate derivative was obtained using this B-type ammonium intercalation compound as an intermediate. Thus, regioselective surface modification of $K_4Nb_6O_{17}{\cdot}3H_2O$ by organophosphonic acid was successfully achieved.

A variety of nanosheets have been obtained by exfoliation of layered materials [52], and various methods have been reported for their exfoliation. A simple method of exfoliation is dispersing layered materials in water. Water molecules can intercalate in the interlayer and promote exfoliation [53]. Bulky ammonium ions intercalated in the interlayer can expand the interlayer distance and decrease interactions between the negative charge and positive charge to cause exfoliation. [54]. Mechanical exfoliation using ultrasonication has also been employed [55]. On the other hand, *in situ* polymerization of organic monomers in the interlayer can also lead to exfoliation of layered materials. A modifier grafted onto the interlayer surface is reacted with monomers and generates polymer chains that expand the interlayer distance. This polymerization method is called the "grafting from" method and is often reported in the field of graphene [56]. Introducing small molecules, such as carboxyl acids, into the interlayer as an initiator group could cause polymerization from the surface of graphene [57, 58]. Another report of the "grafting from" method utilized a layered perovskite, $HLaNb_2O_7{\cdot}xH_2O$, which was modified with organophosphonic acid bearing an initiation group on the interlayer surface, and *N*-isopropylacrylamide (NIPAAm) was polymerized from the initiation group by atom transfer radical polymerization [59]. The interlayer distance was expanded by polymerization, and nanosheets dispersed in water were obtained. Because a thermos-responsive polymer, poly(*N*-isopropylacrylamide), PNIPAAm, was bound to the nanosheet surfaces, the nanosheets were hydrophilic and dispersed in water at below the lower critical solution temperature (LCST). On the contrary, the nanosheets become hydrophobic and aggregate at over the LCST in water.

In this research, the preparation of Janus nanosheets was achieved by taking advantage of the presence of two types of interlayers with different reactivities in $K_4Nb_6O_{17}{\cdot}3H_2O$. Interlayer II of an A-type organophosphonic acid derivative of $K_4Nb_6O_{17}{\cdot}3H_2O$ was reacted with another type of organophosphonic acid. Both sides of niobate nanosheets were modified by two organophosphonic acids regioselectively, because organophosphonic acid could not undergo an exchange reaction and homocondensation. Janus nanosheets could be obtained by exfoliation of the

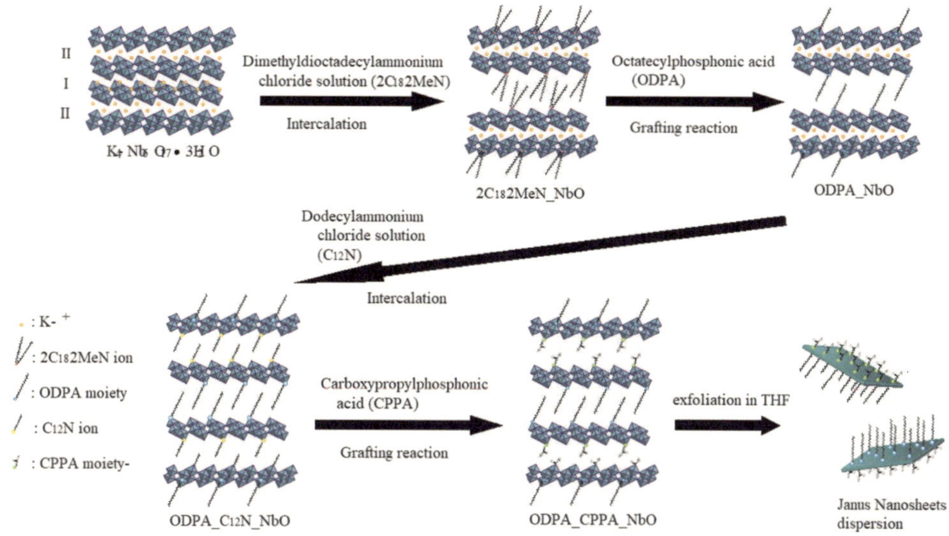

Figure 1.
Preparation of Janus nanosheet.

product into single-layer nanosheets in an appropriate solvent, THF. Lipophilic octadecylphosphonic acid (ODPA) and hydrophilic carboxypropylphosphonic acid (CPPA) were chosen as the organophosphonic acids. The properties of both sides of Janus nanosheets were explored by the AFM phase imaging technique. This report is based on a study first reported in Chemical Communications (**Figure 1**) [60].

2. Experimental section

An A-type alkylammonium intercalation compound, $(2C_{18}2MeN)_{1.0}(K, H)_3[Nb_6O_{17}]$ ($2C_{18}2MeN$ = dioctadecyldimethylammonium ion), was prepared based on the previous report [49]. Octadecylphosphonic acid (ODPA) was synthesized as described elsewhere [61, 62]. Dodecyltrimethylammonium chloride, carboxypropylphosphonic acid (CPPA), 2-butanone, acetone, and tetrahydrofuran (THF) were used without further purification.

First, interlayer I of the A-type alkylammonium intercalation compound was modified by ODPA. The A-type alkylammonium intercalation compound (0.05 g) and ODPA (0.048 g) were used to adjust the Nb:ODPA molar ratio to 1:4 and reacted in 2-butanone (20 mL) at 150°C for 7 days. After the reaction, the crude product was centrifuged, washed with THF and HCl (pH = 3), and air-dried (ODPA_NbO). The cations (K^+, H^+) in interlayer II were then exchanged with the dodecylammonium ion ($C_{12}N^+$) to expand interlayer II. ODPA_NbO (0.1 g) and dodecyltrimethylammonium chloride (0.19 g) were used to adjust the ODPA:$C_{12}N^+$ molar ratio to 1:10 and reacted in water (10 mL) at 80°C for 3 days. After the reaction, the crude product was centrifuged, washed with water, and air-dried (ODPA_C_{12}N_NbO). Then, interlayer II of ODPA_C_{12}N_NbO was modified by CPPA. ODPA_C_{12}N_NbO (0.05 g) and CPPA (0.05 g) were used to adjust the Nb:CPPA molar ratio to 1:10 before reaction in 2-butanone (10 mL) at 80°C for 3 days. After reaction, the crude product was centrifuged, washed with THF and HCl (pH = 3), and corrected as a precipitate (ODPA_CPPA_NbO). After centrifugation, the product dispersed in supernatant was corrected by slow evaporation of THF (ODPA_CPPA_NbO_evaporation). TEM and AFM samples were prepared by stirring ODPA_CPPA_NbO in THF to exfoliate them into single-layer nanosheets.

3. Analysis

XRD analysis (convergence method) was carried out with a Rigaku RINT-1000 diffractometer (Mn-filtered FeKα radiation). XRD analysis (parallel beam method) was performed with a Rigaku SmartLab diffractometer (oblique incidence, FeKα radiation). IR analysis was conducted with a JASCO FT/IR-460 Plus spectrometer by the KBr method. Solid-state ^{31}P magic angle spinning (MAS) NMR spectra were recorded on a JEOL JNM-ECX400 spectrometer. The measurement conditions were as follows: resonance frequency: 160.26 MHz; pulse angle: 90°; pulse delay: 30 s; and MAS frequency: 12 kHz. Triphenylphosphine (−8.4 ppm) was used as a reference. Solid-state ^{13}C cross-polarization (CP)/MAS NMR spectra were recorded on a JEOL JNM-ECX-400 spectrometer. The measurement conditions were as follows: resonance frequency: 99.55 MHz; pulse delay: 5 s; contact time: 1.5 ms; and MAS frequency: 12 kHz. Hexamethylbenzene (17.4 ppm) was used as a reference. ICP-AES measurement was performed using a Thermo Jarrell Ash ICAP-574II instrument. Samples (about 10 mg) were dissolved by heating at 150°C overnight in HF (1 mL), HCl (3 mL), and HNO_3 (4 mL). H_3BO_3 (70 mL) was added as a masking reagent for HF. HF (1 mL), HCl (3 mL), HNO_3 (4 mL), and H_3BO_3 (70 mL) were added to each standard solution for matrix matching. The amounts of C, H, and N in the samples were measured by elemental analysis using a PerkinElmer PE2400II instrument. Transmission electron microscope (TEM) images were observed with a JEOL JEM-1011 microscope operating at 100 kV. A TEM sample was prepared by dropping drops of a dispersion on a Cu 150P grid and drying under reduced pressure. Atomic force microscope (AFM) images were observed with an Agilent 5500 AFM/SPM microscope in the acoustic AC mode under ambient conditions. An ordinary commercial silicon cantilever was used as an AFM tip (e.g., a RTESP-300 from Bruker: resonance frequency ≈ 300 kHz, and spring constant ≈ 40 N/m). Samples for AFM were prepared by spin coating of the dispersion on a Si wafer.

4. Results and discussion

Figure 2 shows ^{13}C CP/MAS NMR spectra of the products. In the spectrum of ODPA_NbO (**Figure 2a**), signals assignable to the octadecyl group were observed. It is likely that the ODPA moiety was introduced into interlayer I, since dioctadecyldimethylammonium ions, whose presence was required for interlayer modification with organophosphonic acids [51], were present only in interlayer I. In the spectrum of ODPA_C_{12}N_NbO (**Figure 2b**), signals assignable to alkyl chains (octadecyl and dodecyl) were observed at 15–43 ppm [51]. In addition, a signal assignable to a carbon atom adjacent to a nitrogen atom was observed at 43 ppm [63], indicating the presence of $C_{12}N^+$. Since $C_{12}N^+$ is known to be intercalated into both interlayer I and interlayer II [48], the intercalation of $C_{12}N^+$ into interlayer II was likely to occur. In the spectrum of ODPA_CPPA_NbO (**Figure 2c**), signals due to alkyl chains were observed at 15–36 ppm. On the other hand, a signal originating from $C_{12}N^+$ at 43 ppm disappeared and a signal due to C=O groups of CPPA was observed at 178 ppm [64]. These results suggest the removal of $C_{12}N^+$ and introduction of the CPPA moiety to ODPA_CPPA_NbO.

Figure 3 shows IR spectra of the products. In the spectrum of ODPA_NbO (**Figure 3a**), absorption bands due to ν (C–H), σ_s (CH_2), and ν (P–O) modes were observed at 2956–2849, 1468, and 1011 cm^{-1}, respectively [65], indicating that ODPA moiety was present in ODPA_NbO. In the spectrum of ODPA_C_{12}N_NbO (**Figure 3b**), an adsorption band at 1540 cm^{-1} assignable to the σ (N–H) mode was observed in addition to the aforementioned adsorption band, indicating that

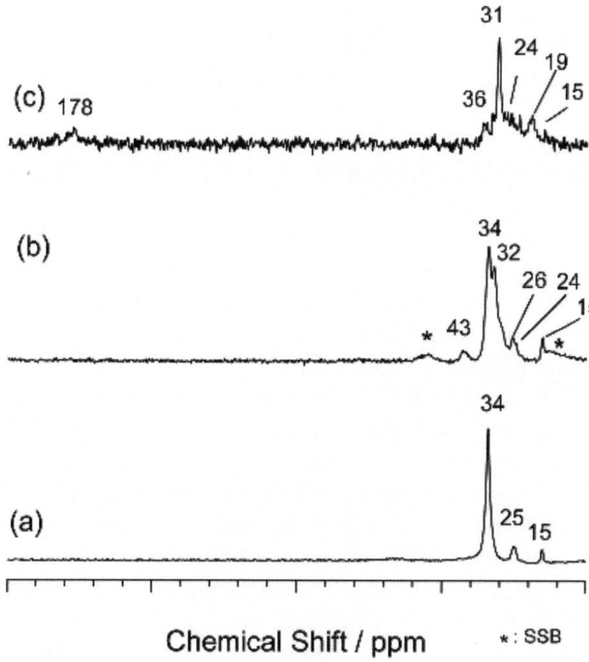

Figure 2.
^{13}C CP/MAS NMR spectra of (a) ODPA_NbO, (b) ODPA_C_{12}N_NbO, and (c) ODPA_CPPA_NbO.

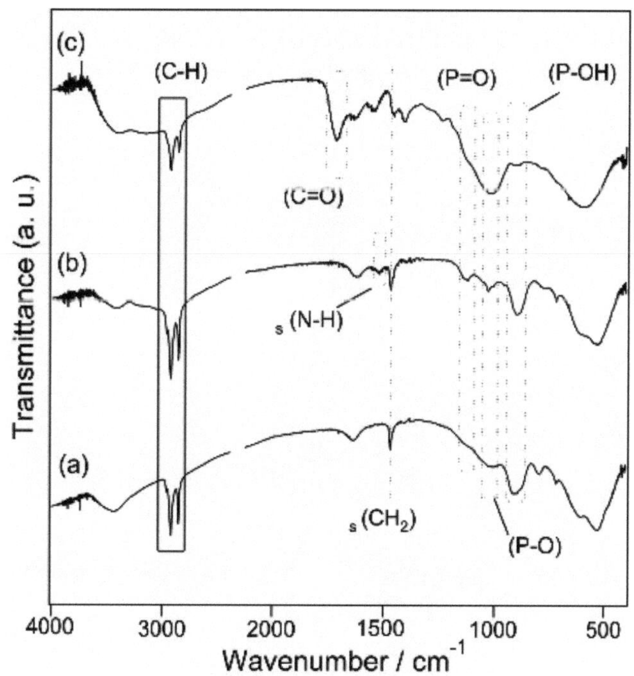

Figure 3.
IR spectra of (a) ODPA_NbO, (b) ODPA_C_{12}N_NbO, and (c) ODPA_CPPA_NbO.

octadecylammonium ions were present in ODPA_C_{12}N_NbO. In the spectrum of ODPA_CPPA_NbO (**Figure 3c**), a new adsorption band that was assignable to the σ (C=O) mode of the CPPA moiety was observed at 1700 cm^{-1} [66], indicating the presence of the CPPA moiety in ODPA_CPPA_NbO. It was reported that an adsorption band due to the ν_{as} (CH_2) of alkyl chain was shifted from 2924.7 cm^{-1} to a lower wavenumber by increasing the packing density of the alkyl chain [67]. In the case of the *all-trans* octadecyl alkyl chain, ν_{as} (CH_2) was observed at 2917.8 cm^{-1} [67, 68] and a σ_s (CH_2) band was observed at 1468 cm^{-1} [68]. In the spectrum of ODPA_NbO, adsorption bands assignable to ν_{as} (CH_2) and ν_s (CH_2) modes were observed at 2918 and 2848 cm^{-1}, respectively, and a σ_s (CH_2) adsorption band was observed at 1468 cm^{-1}. Thus, the alkyl chain in ODPA_NbO was likely to be in an *all-trans* conformation. On the other hand, ν_{as} (CH_2), ν_s (CH_2), and σ_s (CH_2) adsorption bands were observed at 2923, 2852, and 1456 cm^{-1} in the spectrum of ODPA_CPPA_NbO, respectively, indicating that the alkyl chain in ODPA_CPPA_NbO was likely to contain *gauche-blocks*.

Figure 4 shows ^{31}P MAS NMR spectra of the products. A signal was observed at 28 ppm in the spectrum of ODPA_NbO (**Figure 4a**). This signal was shifted upfield from the chemical shift of the ODPA molecule (33 ppm at ^{31}P MAS NMR) by 5 ppm, indicating that interlayer surface modification by ODPA had proceeded and an Nb–O–P bond had been formed [47]. In the spectrum of ODPA_C_{12}N_NbO (**Figure 4b**), a signal was observed at 25 ppm. This signal was shifted upfield from 28 ppm, the chemical shift of ODPA_NbO, by 3 ppm. This shift suggests that $C_{12}N^+$ would change the electronic environment around the P atom by an ion exchange reaction with H^+ of the P–OH group [47], although the details were not yet clarified. Thus, it is likely that $C_{12}N^+$ was intercalated not only in interlayer II, but probably also in interlayer I upon the reaction with ODPA_NbO. In the spectrum of ODPA_CPPA_NbO (**Figure 4c**), a new signal was observed at 31 ppm in addition to the signal at 28 ppm. The signal at 28 ppm was observed in the same position as that of the ODPA moiety of ODPA_NbO, confirming maintenance of the ODPA moiety at interlayer I. Because a signal of a CPPA molecule was observed at 34 ppm, a signal at 31 ppm was assignable to the CPPA moiety. This signal was shifted upfield by 3 ppm, indicating

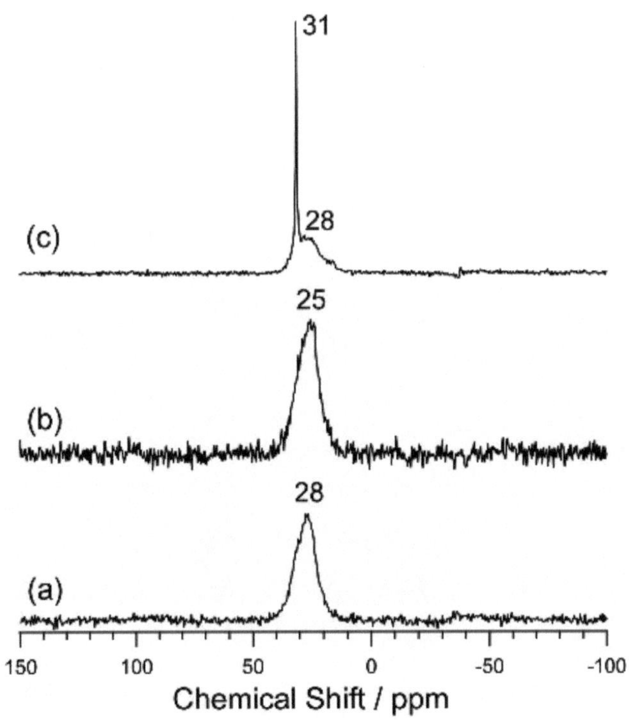

Figure 4.
^{31}P MAS NMR spectra of (a) OPDA_NbO, (b) ODPA_C_{12}N_NbO, and (c) ODPA_CPPA_NbO.

that the CPPA moiety was grafted onto the interlayer surface and a Nb–O–P bond was formed. The above results suggested that ODPA and CPPA formed covalent bonds with the $[Nb_6O_{17}]^{4-}$ sheet surface. Since bands assignable to P–OH groups and P=O groups were observed in the IR spectrum of ODPA_CPPA_NbO, ODPA and CPPA were likely to be in a monodentate environment on the surface of $[Nb_6O_{17}]^{4-}$ sheet.

Table 1 shows the molar ratio calculated from the ICP measurement and elemental analysis. The molar ratio of ODPA_NbO was P:Nb = 1.3:6.0. On the other hand, the molar ratios of ODPA_C_{12}N_NbO and ODPA_CPPA_NbO were P:Nb = 1.3:6.0 and P:Nb = 3.5:6.0, respectively. Intercalation of $C_{12}N^+$ proceeded without release of the ODPA moiety in ODPA_NbO, because the molar ratio of P and Nb of ODPA_NbO did not change after reaction with a dodecylammonium chloride solution. Also, the molar ratio of P to 6 Nb in ODPA_CPPA_NbO increased by 2.2 (3.5 – 1.3), confirming grafting of the CPPA moiety. Assuming Nb = 6.0, the maximum modification amounts for interlayer I and II are 2.0 [51]. Since the Nb–O–P bond was stable with respect to hydrolysis and no homocondensation between two P–OH groups of phosphonic acid occurred under mild conditions [36], the amount of the ODPA moiety in interlayer I was estimated to be 1.3 (65% of the maximum modification amount), that of the CPPA moiety at interlayer I was in the range of 0.2–0.7 (10–35% of maximum modification amount), and that of the CPPA moiety in interlayer II was in the range of 1.5–2.0 (75–100% of the maximum modification amount). Thus, an organic derivative with interlayer I and interlayer II dominantly modified with hydrophobic ODPA and hydrophilic CPPA, respectively, were successfully prepared (**Figure 5**).

Based on the nitrogen ratio of ODPA_NbO, it seems that a small amount of unreacted A-type alkylammonium intercalation compound was present in ODPA_NbO or a small number of released $2C_{18}2MeN$ ions were present in interlayer I *via* ion exchange. Since the amount of K^+ in interlayer II in ODPA_C_{12}N_NbO decreased, an ion exchange reaction between K^+ and H^+ ions at interlayer II and $C_{12}N^+$ proceeded. Because no nitrogen was detected in ODPA_CPPA_NbO, $C_{12}N^+$ was completely removed from interlayer I and II after the reaction with CPPA.

A THF dispersion of nanosheets was easily attained by dispersing ODPA_CPPA_NbO in THF. The resulting dispersion was cast on a TEM grid, and TEM observation was carried out (**Figure 6**). A sheet-like morphology with low contrast was observed. Spots observed in the electron diffraction (ED) pattern can be assigned to 200, 202, and 002 of the orthorhombic cells, and the lattice parameters were calculated to be a = 0.80 nm and c = 0.64 nm. This ED pattern was thus a b-axis incidence pattern of $K_4Nb_6O_{17}$·$3H_2O$ [69]. Based on these results, ODPA_CPPA_NbO was synthesized while maintaining the crystal structure of the $[Nb_6O_{17}]^{4-}$ nanosheets.

Figure 7 shows XRD patterns of the products. The d values of low-angle diffractions due to repeating distances were as follows: the d value of ODPA_NbO, A-type derivative (**Figure 7a**), was 5.67 nm and the d value of ODPA_C_{12}N_NbO (**Figure 7b**) was 4.03 nm. If intercalation of $C_{12}N^+$ into interlayer II proceeded while maintaining an A-type stacking sequence, the d value of ODPA_C_{12}N_NbO is likely to have increased from that of ODPA_NbO. It is possible that a B-type stacking sequence was generated due to exfoliation and restacking during the reaction,

	Nb/–	K/–	P/–	N/–
ODPA_NbO	6.0	2.6	1.3	0.082
ODPA_C_{12}N_NbO	6.0	0.58	1.3	1.8
ODPA_CPPA_NbO	6.0	0.49	3.5	—

Table 1.
Molar ratios of ODPA_NbO, ODPA_C_{12}N_NbO, and ODPA_CPPA_NbO.

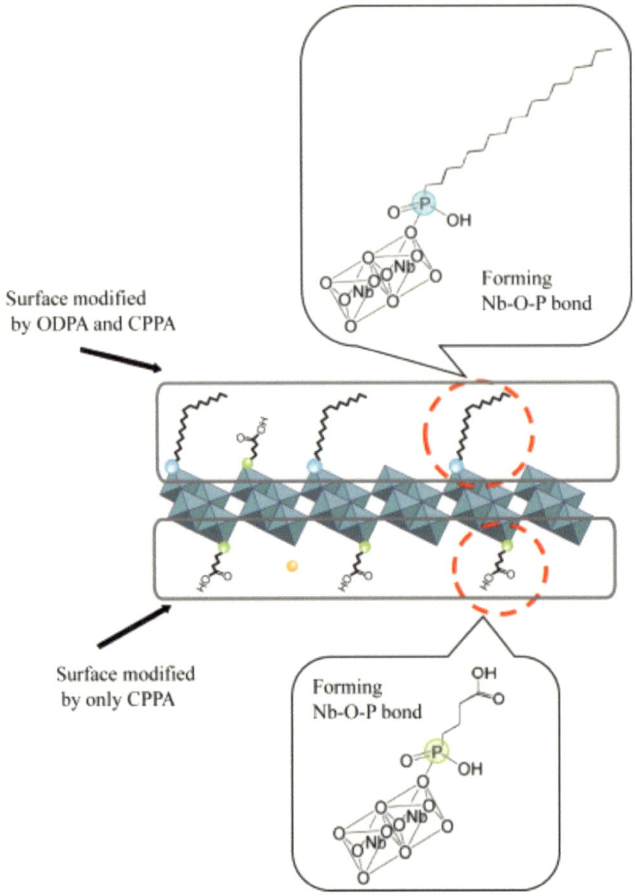

Figure 5.
Proposed structure of ODPA_CPPA_NbO.

resulting in a smaller repeating distance. Also, the *d* values of ODPA_CPPA_NbO (**Figure 7c**) and ODPA_CPPA_NbO_evaporation (**Figure 7d**) were 2.41 and 4.74 nm, respectively. The stacking sequence would therefore be changed by reaction between ODPA_C_{12}N_NbO and CPPA.

Here, the difference between these two *d* values is discussed. If ODPA_CPPA_NbO is a B-type derivative, the thickness of an organic moiety layer (sum of an ODPA monolayer and a CPPA monolayer) can be calculated by subtracting 0.82 nm, the niobate layer thickness, from 2.41 nm to make 1.59 nm [51]. The repeating distance of an A-type derivative could thus be estimated as the sum of a double niobate layer thickness and a double organic layer thickness. The repeating distance of an A-type derivative can therefore be estimated as follows: (1.59 nm × 2) + (0.82 nm × 2) = 4.82 nm. This value is approximately equal to *d* = 4.74 nm of ODPA_CPPA_NbO_evaporation. From these estimations, it is proposed that ODPA_CPPA_NbO is a B-type derivative and ODPA_CPPA_NbO_evaporation is an A-type derivative. As shown in **Figure 8**, a B-type derivative could be generated by forced restacking *via* centrifugation of exfoliated nanosheets (**Figure 8a** and **b**). On the other hand, an A-type derivative, in which hydrophilic groups faced each other and hydrophobic groups faced each other, was obtained by slow evaporation under mild conditions (**Figure 8c**).

The crystallite sizes calculated from diffraction of the repeating distances using Scherrer's formula were 3.67 and 7.71 nm for ODPA_CPPA_NbO and

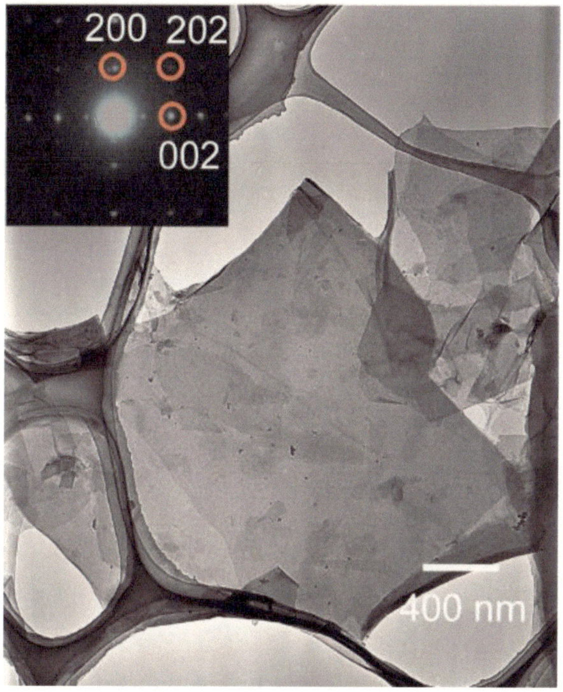

Figure 6.
TEM image of exfoliated ODPA_CPPA_NbO. The inset shows the corresponding ED pattern.

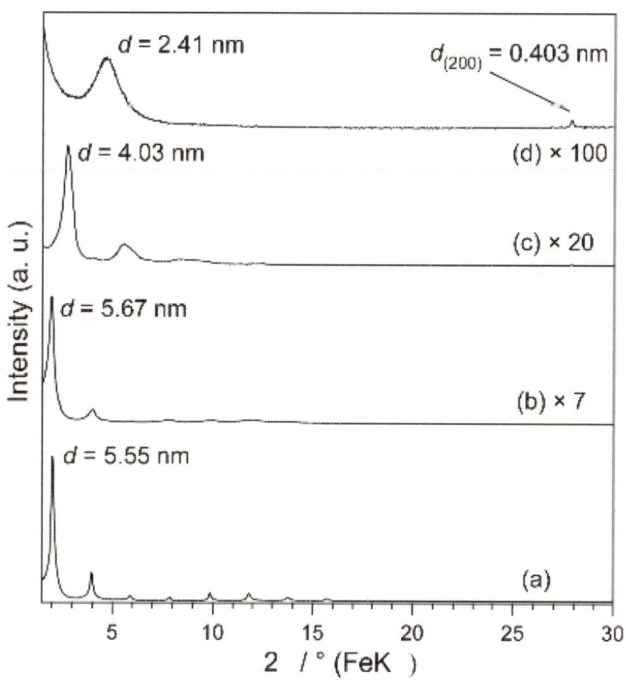

Figure 7.
XRD patterns of (a) ODPA_NbO, (b) ODPA_C_{12}N_NbO, (c) ODPA_CPPA_NbO, and (d) ODPA_CPPA_NbO_evaporation.

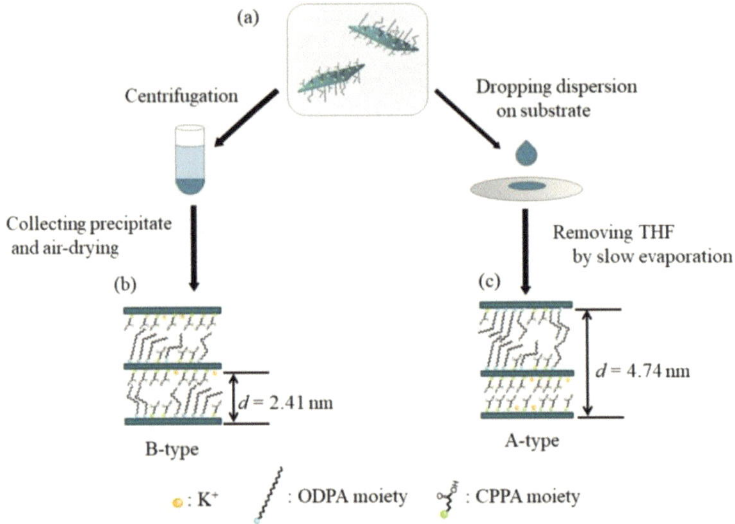

Figure 8.
The estimated structures of ODPA_CPPA_NbO: Possible routes from (a) ODPA_C12N_NbO to (b) ODPA_CPPA_NbO and (c) ODPA_CPPA_NbO_evaporation.

ODPA_CPPA_NbO_evaporation, respectively. The crystallite size of ODPA_CPPA_NbO_evaporation was larger than that of ODPA_CPPA_NbO. It should be noted that underestimation could occur with use of lowest-angle diffractions due to the presence of strain [32]. The crystallite sizes could therefore reflect the average thickness of the particles in the stacking direction, making the number of stacked ODPA_CPPA_NbO nanosheets lower than that of stacked ODPA_CPPA_NbO_evaporation nanosheets. On the other hand, the estimated crystallite size could be interpreted as average thickness of a portion of the stacked sheets with an A-type or B-type stacking sequence. Based on this interpretation, ODPA_CPAN_NbO formed via forced restacking by centrifugation has lower stacking regularity or more random stacking than ODPA_CPPA_NbO_evaporation nanosheets restacked under mild conditions.

Figure 9 shows an AFM image of a sample prepared by spin coating of a THF dispersion of ODPA_CPPA_NbO on a Si wafer. It contained many nanosheets that showed a relatively uniform thickness in the range of 2.5–3.0 nm (**Figure 9A**). This thickness range is approximately equal to the d value of B-type ODPA_CPPA_NbO, indicating that ODPA_CPPA_NbO was exfoliated into single-layer nanosheets that were casted on a Si wafer.

As marked by the a and b arrows in **Figure 9**, two different colored nanosheet surfaces were observed in the phase image (**Figure 9B**). This indicates the presence of two different faces (36–37° and 38–39°) in each Janus nanosheet. This phase difference in the AFM phase image corresponds to the tapping phase gap in the vibration amplitude, and it was reported that the phase difference could occur with a difference in the crystallinity, viscosity, and adhesion of the sample surface [70, 71].

The Janus nanosheets consisted of a hydrophobic surface, which was dominantly covered with the ODPA moiety, and a hydrophilic surface, which was modified with the CPPA moiety. As a result, two chemically different surfaces gave different phases due to differences in the interactions between the apex of the AFM probe and the surfaces of the nanosheets and distinguished visually in the phase image. The origin of the phase contrast would be due to differences in viscosity and hydrophilicity/hydrophobicity. Since the apex of the AFM probe used in this measurement was hydrophilic, it is likely that the high-phase surface and a low-phase surface were assignable to the hydrophilic CPPA moiety and hydrophobic ODPA moiety, respectively.

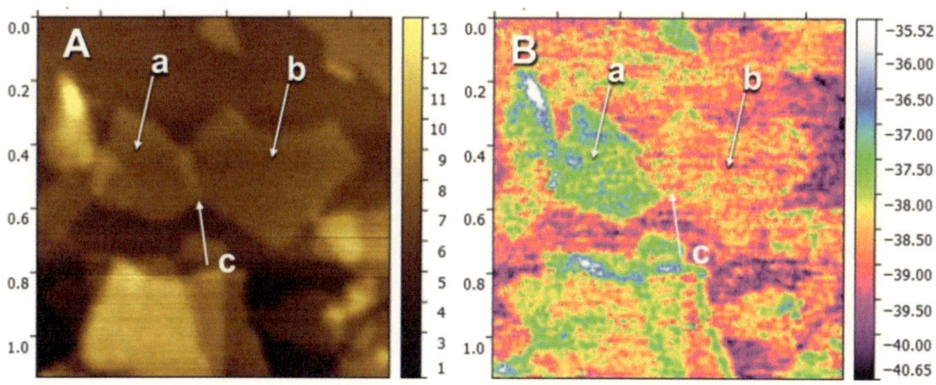

Figure 9.
Topographic (A) and phase (B) AFM images of ODPA_CPPA_NbO Janus nanosheets.

The c arrow in **Figure 9** marks the overlapping area of two nanosheets (a and b). Obviously, these area possessed double-layer thickness. The color of the phase image of this area (**Figure 9B**) indicates that nanosheet b partially overlapped nanosheet a. Thus, these results indicate that hydrophilic and lipophilic surfaces are facing each other. These results also indicate that the nanosheets prepared in this study exhibited hydrophobicity on one side and hydrophilicity on the other.

5. Conclusions

Janus nanosheets were successfully prepared by regioselective and sequential surface modification and exfoliation of $K_4Nb_6O_{17}·3H_2O$, whose interlayer I and interlayer II were dominantly modified by ODPA and CPPA, respectively. Since organophosphonic acids bearing various functional groups can be easily synthesized, Janus nanosheet surfaces can exhibit various properties in addition to hydrophobicity and hydrophilicity. The Janus nanosheets prepared by the present method can be dispersed in many solvents, moreover, because organophosphonic moieties are bound to niobate nanosheets by covalent bonds. The Janus nanosheets prepared in this study can be expected to be applied in surface chemistry research because of the hydrophobicity and hydrophilicity on opposing sides of the nanosheets. Also, by changing the functional groups of organophosphonic acids, novel two-dimensional materials with various functions with potential applications in various fields can be realized.

Acknowledgements

This work was financially supported in part by a Grant-in-Aid for Scientific Research on Innovative Areas "New Polymeric Materials Based on Element-Blocks (No. 2401)" (JSPS KAKENHI Grant Numbers JP24102002), "Coordination Asymmetry (JP 23655205)," and Grant-in-Aid for Challenging Exploratory Research (JP17H05378). Reproduced from Ref. [60] with permission from the Royal Society of Chemistry.

Conflict of interest

There are no conflicts to declare.

Author details

Ryoko Suzuki[1], Mitsuhito Sudo[1], Megumi Hirano[2], Naokazu Idota[3], Masashi Kunitake[4], Taisei Nishimi[5] and Yoshiyuki Sugahara[1,6*]

1 Department of Applied Chemistry, School of Advanced Science and Engineering, Waseda University, Tokyo, Japan

2 Technical Division, Faculty of Engineering, Kumamoto University, Kumamoto, Japan

3 Department of Chemical Science and Technology, Faculty of Bioscience and Applied Chemistry, Hosei University, Tokyo, Japan

4 Faculty of Advanced Science and Technology, Kumamoto University, Kumamoto, Japan

5 Japan Technological Research Association of Artificial Photosynthetic Chemical Process (ARPChem), Tokyo, Japan

6 Kagami Memorial Research Institute for Science and Technology, Waseda University, Tokyo, Japan

*Address all correspondence to: ys6546@y.waseda.jp

References

[1] de Gennes PG. Soft matter (Nobel lecture). Angewandte Chemie, International Edition. 1992;**31**:842-845. DOI: 10.1002/anie.199208421

[2] Nishisako T, Torii T, Takahashi T, Takizawa Y. Synthesis of monodisperse bicolored Janus particles with electrical anisotropy using a microfluidic co-flow system. Advanced Materials. 2006;**18**:1152-1156. DOI: 10.1002/adma.200502431

[3] Binks BP, Fletcher PDI. Particles adsorbed at the oil–water Interface: A theoretical comparison between spheres of uniform wettability and "Janus" particles. Langmuir. 2001;**17**:4708-4710. DOI: 10.1021/la0103315

[4] Kim M, Anthony SA, Granick S. Activated surface diffusion in a simple colloid system. Physical Review Letters. 2009;**102**:178303. DOI: 10.1103/PhysRevLett.102.178303

[5] Seo KD, Doh J, Kim DS. One-step microfluidic synthesis of Janus microhydrogels with anisotropic thermo-responsive behavior and organophilic/hydrophilic loading capability. Langmuir. 2013;**29**:15137-15141. DOI: 10.1021/la403015y

[6] Walther A, Muller AHE. Janus particles: Synthesis, self-assembly, physical properties, and applications. Chemical Reviews. 2013;**113**:5194-5261. DOI: 10.1021/cr300089t

[7] Lattuada M, Hatton TA. Synthesis, properties and applications of Janus nanoparticles. Nano Today. 2011;**6**:286-308. DOI: 10.1016/j.nantod.2011.04.008

[8] Jiang S, Chen Q, Tripathy M, Luijten E, Schweizer KS, Granick S. Janus particle synthesis and assembly. Advanced Materials. 2010;**22**:1060-1071. DOI: 10.1002/adma.200904094

[9] Erhardt R, Zhang M, Boker A, Zettl H, Abetz C, Frederik P, et al. Amphiphilic Janus micelles with polystyrene and poly(methacrylic acid) hemispheres. Journal of the American Chemical Society. 2003;**125**:3260-3267. DOI: 10.1021/ja028982q

[10] Qi H, Wang W, Li CY. Janus polymer single crystal nanosheet via evaporative crystallization. ACS Macro Letters. 2014;**3**:675-678. DOI: 10.1021/mz5002806

[11] Perro A, Reculusa S, Pereira F, Delville MH, Mingotaud C, Duguet E, et al. Towards large amounts of Janus nanoparticles through a protection–deprotection route. Chemical Communications. 2005;(44):5542-5543. DOI: 10.1039/B507486J

[12] Hawker CJ. "Living" free radical polymerization: A unique technique for the preparation of controlled macromolecular architectures. Accounts of Chemical Research. 1997;**30**:373-382. DOI: 10.1021/ar960248m

[13] Zhao B, Zhou H, Liu C, Long Y, Yang G, Tungd CH, et al. Fabrication and directed assembly of magnetic Janus rods. New Journal of Chemistry. 2016;**40**:6541-6545. DOI: 10.1039/C6NJ00825A

[14] Oratis AT, Farmer TP, Bird JC. Capillary induced twisting of Janus cylinders. Soft Matter. 2017;**13**:7556-7561. DOI: 10.1039/C7SM01288H

[15] de Leon AC, Rodier BJ, Luo Q, Hemmingsen CM, Wei P, Abbasi K, et al. Distinct chemical and physical properties of Janus nanosheets. ACS Nano. 2017;**11**:7485-7493. DOI: 10.1021/acsnano.7b04020

[16] Stupp SI, Son S, Lin HC, Li LS. Synthesis of two-dimensional polymers. Science. 1993;**259**:59-63. DOI: 10.1126/science.259.5091.59

[17] Walther A, Andre X, Drechsler M, Abetz V, Muller AHE. Janus Discs. Journal of the American Chemical Society. 2007;**129**:6187-6198. DOI: 10.1021/ja068153v

[18] Walther A, Drechsler M, Muller AHE. Structures of amphiphilic Janus discs in aqueous media. Soft Matter. 2009;**5**:385-390. DOI: 10.1039/B812321G

[19] Liang F, Shen K, Qu X, Zhang C, Wang Q, Li J, et al. Inorganic Janus nanosheets. Angewandte Chemie, International Edition. 2011;**50**: 2379-2382. DOI: 10.1002/anie.201007519

[20] Yang H, Liang F, Wang X, Chen Y, Zhang C, Wang Q, et al. Responsive Janus composite nanosheets. Macromolecules. 2013;**46**:2754-2759. DOI: 10.1021/ma400261y

[21] Liu Y, Liang F, Wang Q, Qu X, Yang Z. Flexible responsive Janus nanosheets. Chemical Communications. 2015;**51**:3562-3565. DOI: 10.1039/C4CC08420A

[22] Stçter M, Gçdrich S, Feicht P, Rosenfeldt S, Thurn H, Neubauer JW, et al. Controlled exfoliation of layered silicate heterostructures into bilayers and their conversion into giant Janus platelets. Angewandte Chemie, International Edition. 2016;**55**:7398-7402. DOI: 10.1002/anie.201601611

[23] Ng SW, Noor N, Zheng Z. Graphene-based two-dimensional Janus materials. NPG Asia Materials. 2018;**10**:217-237. DOI: 10.1038/s41427-018-0023-8

[24] Zhang L, Yu J, Yang M, Xie Q, Peng H, Liu Z. Janus graphene from asymmetric two-dimensional chemistry. Nature Communications. 2013;**4**:1443. DOI: 10.1038/ncomms2464

[25] Wu H, Yi W, Chen Z, Wang H, Du Q. Janus graphene oxide nanosheets prepared via Pickering emulsion template. Carbon. 2015;**93**:473-483. DOI: 10.1016/j.carbon.2015.05.083

[26] Osada M, Sasaki T. Exfoliated oxide nanosheets: New solution to nanoelectronics. Journal of Materials Chemistry. 2009;**19**:2503-2511. DOI: 10.1039/B820160A

[27] Joussein E, Petit S, Churchman J, Theng B, Righi D, Delvaux B. Halloysite clay minerals—A review. Clay Minerals. 2005;**40**(04):383-426. DOI: 10.1180/0009855054040180

[28] Madejova J. FTIR techniques in clay mineral studies. Vibrational Spectroscopy. 2003;**31**(1):1-10. DOI: 10.1016/S0924-2031(02)00065-6

[29] Kosuge K, Yamazaki A, Tsunashima A, Otsuka R. Hydrothermal synthesis of magadiite and kenyaite. Journal of the Ceramic Society of Japan. 1992;**100**(3):326-331. DOI: 10.2109/jcersj.100.326

[30] Brindley GW. Unit cell of magadiite in air, in vacuo, and under other conditions. American Mineralogist. 1969;**54**(11-1):1583-1591

[31] Wolf F, Schwieger W. Ion exchange of monovalent cations in synthetic sodium polysilicates with layer structure. Zeitschrift für Anorganische und Allgemeine Chemie. 1979;**457**(1):224-228. DOI: 10.1002/zaac.19794570128

[32] Sasaki T, Watanabe M. Osmotic swelling to exfoliation. Exceptionally high degrees of hydration of a layered titanate. Journal of the American Chemical Society. 1998;**120**(19): 4682-4689. DOI: 10.1021/ja974262l

[33] Bizeto MA, Shiguihara AL, Constantino VRL. Layered niobate nanosheets: Building blocks for advanced materials assembly. Journal of Materials Chemistry. 2009;**19**: 2512-2525. DOI: 10.1039/B821435B

[34] Osada M, Sasaki T. Two-dimensional dielectric nanosheets: Novel nanoelectronics from nanocrystal building blocks. Advanced Materials. 2012;**24**:210-228. DOI: 10.1002/adma.201103241

[35] Guerrero G, Alauzun JG, Granier M, Laurencin D, Mutin PH. Phosphonate coupling molecules for the control of surface/interface properties and the synthesis of nanomaterials. Dalton Transactions. 2013;**42**:12569-12585. DOI: 10.1039/c3dt51193f

[36] Guerrero G, Mutin PH, Vioux A. Anchoring of phosphonate and phosphinate coupling molecules on titania particles. Chemistry of Materials. 2001;**13**:4367-4373. DOI: 10.1021/cm001253u

[37] Takahashi N, Kuroda K. Materials design of layered silicates through covalent modification of interlayer surfaces. Journal of Materials Chemistry. 2011;**21**:14336-14353. DOI: 10.1039/C1JM10460H

[38] Yanagisawa T, Kuroda K, Kato C. Organic derivatives of layered polysilicates. II. Reaction of magadiite and kenyaite with diphenylmethylchlorosilane. Bulletin of the Chemical Society of Japan. 1988;**61**:3743-3745. DOI: 10.1246/bcsj.61.3743

[39] Fujita I, Kuroda K, Ogawa M. Synthesis of interlamellar silylated derivatives of magadiite and the adsorption behavior for aliphatic alcohols. Chemistry of Materials. 2003;**15**:3134-3141. DOI: 10.1021/cm011698y

[40] Mochizuki D, Shimojima A, Kuroda K. Formation of a new crystalline silicate structure by grafting dialkoxysilyl groups on layered octosilicate. Journal of the American Chemical Society. 2002;**124**: 12082-12083. DOI: 10.1021/ja027512t

[41] Ide Y, Ogawa M. Surface modification of a layered alkali titanate with organosilanes. Chemical Communications. 2003:1262-1263. DOI: 10.1039/B301222K

[42] Nakato T, Hashimoto S. Dispersion of layered hexaniobate in organic solvents through silylation and liquid crystalline behavior of the colloidal suspension. Chemistry Letters. 2007;**36**:1240-1241. DOI: 10.1246/cl.2007.1240

[43] Mercier L, Facey GA, Detellier C. Organo-layered silicates. Interlamellar intercation and grafting of ethylene glycol in magadiite. Journal of the Chemical Society, Chemical Communications. 1994:2111-2112. DOI: 10.1039/C39940002111

[44] Mitamura Y, Komori Y, Hayashi S, Sugahara Y, Kuroda K. Interlamellar esterification of H-magadiite with aliphatic alcohols. Chemistry of Materials. 2001;**13**:3747-3753. DOI: 10.1021/cm010029h

[45] Tahara S, Ichikawa T, Kajiwara G, Sugahara Y. Reactivity of the Ruddlesden-Popper phase $H_2La_2Ti_3O_{10}$ with organic compounds: Intercalation and grafting reactions. Chemistry of Materials. 2007;**19**:2352-2358. DOI: 10.1021/cm0623662

[46] Asai Y, Ariake Y, Saito H, Idota N, Matsukawa K, Nishino T, et al. Layered perovskite nanosheets bearing fluoroalkoxy groups: Their preparation and application in epoxy-based hybrids. RSC Advances. 2014;**4**:26932-26939. DOI: 10.1039/c4ra01777c

[47] Shimada A, Yoneyama Y, Tahara S, Mutin PH, Sugahara Y. Interlayer surface modification of the protonated ion-exchangeable layered perovskite $HLaNb_2O_7$·xH_2O with organophosphonic acids. Chemistry of Materials. 2009;**21**:4155-4162. DOI: 10.1021/cm900228c

[48] Nakato T, Kuroda K, Kato C. Syntheses of intercalation compounds of layered niobates with methylviologen and their photochemical behavior. Chemistry of Materials. 1992;**4**:128-132. DOI: 10.1021/cm00019a027

[49] Nakato T, Kameyama M, Wei Q, Haga J. Structural response of organically modified layered niobate $K_4Nb_6O_{17}$ to the adsorption of 2,4-dichlorophenol. Microporous and Mesoporous Materials. 2008;**110**:223-231. DOI: 10.1016/j.micromeso.2007.06.011

[50] Wei Q, Nakato T. Competitive adsorption of phenols on organically modified layered hexaniobate $K_4Nb_6O_{17}$. Microporous and Mesoporous Materials. 2006;**96**:84-91. DOI: 10.1016/j.micromeso.2006.06.028

[51] Kimura N, Kato Y, Suzuki R, Shimada A, Tahara S, Nakato T, et al. Single- and double-layered organically modified nanosheets by selective interlayer grafting and exfoliation of layered potassium hexaniobate. Langmuir. 2014;**30**:1169-1175. DOI: 10.1021/la404223x

[52] Nicolosi V, Chhowalla M, Kanatzidis MG, Strano MS, Coleman JN. Liquid exfoliation of layered materials. Science. 2013;**340**(6139):1226400-1226419. DOI: 10.1126/science.1226419

[53] Anderson RL, Ratcliffe I, Greenwell HC, Williams PA, Cliffe S, Coveney PV. Clay swelling—A challenge in the oilfield. Earth Science Reviews. 2010;**98**(3-4):201-216. DOI: 10.1016/j.earscirev.2009.11.003

[54] Ma RZ, Sasaki T. Nanosheets of oxides and hydroxides: Ultimate 2D charge-bearing functional crystallites. Advanced Materials. 2010;**22**(45):5082-5104. DOI: 10.1002/adma.201001722

[55] Fan X, Xu P, Zhou D, Sun Y, Li YC, Nguyen MAT, et al. Fast and efficient preparation of exfoliated 2H MoS_2 nanosheets by sonication-assisted Lithium intercalation and infrared laser-induced 1T to 2H phase reversion. Nano Letters. 2015;**15**(9):5956-5960. DOI: 10.1021/acs.nanolett.5b02091

[56] Layek RK, Nandi AK. A review on synthesis and properties of polymer functionalized graphene. Polymer. 2013;**54**(19):5087-5103. DOI: 10.1016/j.polymer.2013.06.027

[57] Liu J, Chen C, He C, Zhao J, Yang X, Wang H. Synthesis of graphene peroxide and its application in fabricating super extensible and highly resilient nanocomposite hydrogels. ACS Nano. 2012;**6**:8194-8208. DOI: 10.1021/nn302874v

[58] Fang M, Wang K, Lu H, Yang Y, Nutt S. Single-layer graphene nanosheets with controlled grafting of polymer chains. Journal of Materials Chemistry. 2010;**20**:1982-1992. DOI: 10.1039/B919078C

[59] Idota N, Fukuda S, Tsukahara T, Sugahara Y. Preparation of thermoresponsive nanosheets exhibiting phase transitions in water via surface modification of layered perovskite nanosheets with poly(*N*-isopropylacrylamide) (PNIPAAm). Chemistry Letters. 2015;**44**(2):203-205. DOI: 10.1246/cl.140956

[60] Suzuki R, Sudo M, Hirano M, Idota N, Kunitake M, Nishimi T, et al. Inorganic Janus nanosheets bearing two types of covalently bound organophosphonate groups via regioselective surface modification of $K_4Nb_6O_{17}{\cdot}3H_2O$. Chemical Communications. 2018;**54**:5756-5759. DOI: 10.1039/C8CC02892C

[61] Woodward JT, Ulman A, Schwartz DK. Self-assembled monolayer growth of octadecylphosphonic acid on mica.

Langmuir. 1996;**12**:3626-3629. DOI: 10.1021/la9510689

[62] Kim BY, Ratcliff EL, Armstrong NR, Kowalewski T, Pyun J. Ferrocene functional polymer brushes on indium tin oxide *via* surface-initiated atom transfer radical polymerization. Langmuir. 2010;**26**:2083-2092. DOI: 10.1021/la902590u

[63] Mello MR, Phanon D, Silveira GQ, Llewellyn PL, Ronconi CM. Amine-modified MCM-41 mesoporous silica for carbon dioxide capture. Microporous and Mesoporous Materials. 2011;**143**:174-179. DOI: 10.1016/j.micromeso.2011.02.022

[64] Ghosh S, Yu WZ, Kang S, Bhowmik PC, Xing BS. Sorption and fractionation of a peat derived humic acid by kaolinite, montmorillonite, and goethite. Pedosphere. 2009;**19**(1):21-30. DOI: 10.1016/S1002-0160(08)60080-6

[65] Shurvell HF. Spectra-structure correlations in the mid- and far-infrared. In: Chalmers JM, Griffiths PR, editors. Handbook of Vibrational Spectroscopy, Sample Characterization and Spectral Data Processing. Vol. 3. Chichester: John Wiley & Sons; 2002. pp. 1783-1816. DOI: 10.1002/0470027320

[66] Katsumoto Y, Tanaka T, Sato H, Ozaki Y. Conformational change of poly(*N*-isopropylacrylamide) during the coil-globule transition investigated by attenuated total reflection/infrared spectroscopy and density functional theory calculation. The Journal of Physical Chemistry. A. 2002;**106**: 3429-3435. DOI: 10.1021/jp0124903

[67] Quinones R, Rodriguez K, Iuliucci RJ. Investigation of phosphonic acid surface modifications on zinc oxide nanoparticles under ambient conditions. Thin Solid Films. 2014;**565**:155-164. DOI: 10.1016/j.tsf.2014.06.057

[68] Vaia RA, Teukolsky RK, Giannelis EP. Interlayer structure and molecular environment of alkylammonium layered silicates. Chemistry of Materials. 1994;**6**:1017-1022. DOI: 10.1021/cm00043a025

[69] Nassau K, Shiever W, Bernstein JL. Crystal growth and properties of mica-like potassium niobates. Journal of the Electrochemical Society. 1969;**116**:348-353. DOI: 10.1149/1.2411844

[70] Ricci D, Braga C. Imaging methods in atomic force microscopy. In: Ricci D, Braga C, editors. Atomic Force Microscopy Biomedical Methods and Applications. New York City: Humana Press; 2014. DOI: P13-23. DOI:10.1385/1592596479

[71] Magonov SN, Whangbo MH. Phase imaging and stiffness in tapping-mode atomic force microscopy. Surface Science. 1997;**375**:L385-L391. DOI: 10.1016/S0039-6028(96)01591-9

Section 2

Functional Magnetic Nanomaterials

Chapter 4

Intriguing Properties and Applications of Functional Magnetic Materials

Sabah M. Abdelbasir and Ahmed Esmail Shalan

Abstract

Functional magnetic materials, also called the smart materials of the future, are a group of materials having important and interesting physical properties, which can be affected when an external magnetic field is applied. They are intriguing models that have a strong impact on improving different technologies. Their magnetic response to an external magnetic field can be represented as paramagnetic, diamagnetic, ferromagnetic, or antiferromagnetic. Compared with bulk materials, they show uncommon magnetic behavior as a result of their surface/interface effects, electronic charge transfer, and magnetic interactions. They can be used in different vital applications like data storage systems, refrigeration, magnetic recording, and medical studies. In this chapter, essential attractive magnetic properties that are relevant to different applications will be explained.

Keywords: functional materials, magnetic nanoparticles, magnetic field, physical properties, applications

1. Introduction

1.1 Magnetic materials

Magnetic materials play a crucial role in the progress of industrial development and scientific growth. They are constantly used in power generation and transmission, electronic devices, analog and digital data storage, medical devices, magnetic therapy and drug delivery, sensors and scientific equipment, etc. Functional magnetic materials are materials with unique physical properties, which can be affected when subjected to an applied excitement such as magnetic field. They are considered as the smart materials of the future. A material can be applied in magnetic refrigerators when a change in the entropy across its magnetic ordering temperature occurs. This functionality of a magnetic material has huge possibility to be used as an alternative cooling technology and it is based on magnetocaloric effect (MCE), which is reversible temperature change in a magnetic material when a variable magnetic field is applied. This functionality additionally offers the prospect of a compact, highly efficient, and environment-friendly alternative to the most commonly used vapor-compression-based freezing system. The main challenges are the availability of high magnetocaloric materials in large quantities exhibiting large MCE at room temperature in a reasonable magnetic field as well as low hysteretic losses.

Magnetic nanoparticles have been the focus of research because of their interesting properties, which doubtless may see use in data storage and processing, spintronics, catalysis, drug delivery, magnetic resonance imaging (MRI), environmental studies, etc. These materials show uncommon magnetic behavior compared with bulk materials, principally because of their surface/interface effects, electronic charge transfer, and magnetic interactions. The local magnetic properties with the size scale of nanometers play the key role in the microstructure-magnetic properties interplay in permanent magnets as **Figure 1** illustrates. The typical phenomena related to nanoscale structures are the increased relevance of surface effects, defects, and the existence of new phases. Therefore, these phenomena can be utilized in developing new magnetic nanoparticles.

1.1.1 Hard magnetic materials

Several permanent magnet materials were discovered within the past century. Techniques to effectively manufacture these magnets have been shown [2]. Device designs using such magnets in different active and inactive applications have been fruitfully exploited. The energy product of permanent magnets has been improved, commencing from ≈1 MGOe for steels, increasing to ≈3 MGOe for hexagonal ferrites, and finally peaking at ≈56 MGOe for neodymium-iron boron magnets during the previous few years. With this, almost 90% of the limit for the energy density, $(BH)_{max}$, (based on the Nd_2Fe_{14} B phase) can be attained in commercially produced sintered Nd-Fe-B grades. The historic development, spanning about 100 years, of such permanent magnets is shown in **Figure 2**.

However, the search for novel hard magnetic compounds with higher remnant magnetization has, to some extent, settled and no more breakthrough is noticeable. On the other side, only a modest number of ternary and quaternary systems have been explored as yet. The approach of nanocomposites is currently the most actively chased as well as exchange-coupled with a soft magnetic phase, which has

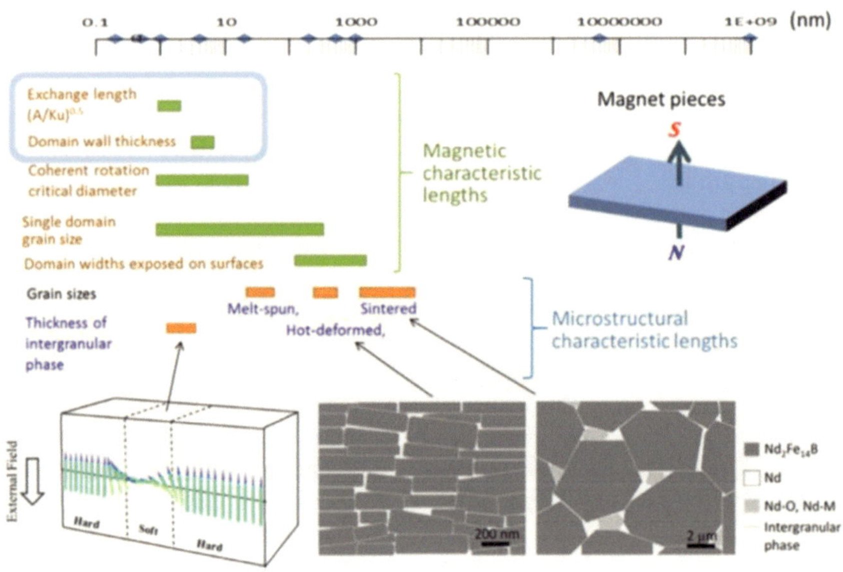

Figure 1.
Magnetic characteristic lengths and illustration of typical microstructures in permanent magnets [1]. Reproducibility with permission from IOP publisher.

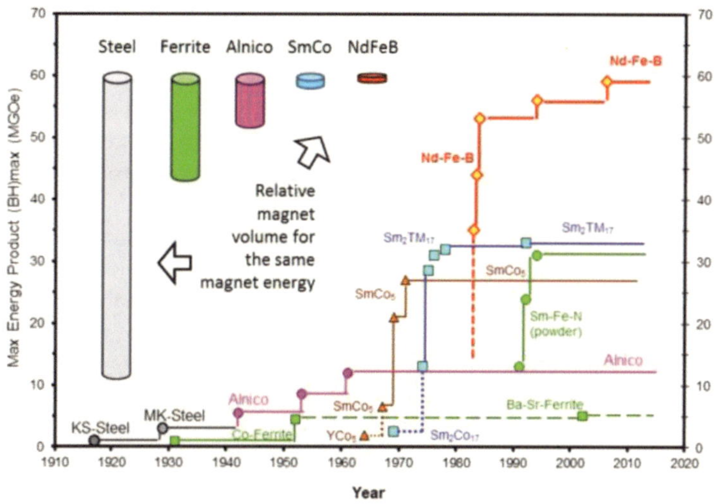

Figure 2.
Development in the energy density $(BH)_{max}$ at room temperature of hard magnetic materials in the twentieth century and presentation of different types of materials with comparable energy density. Reproduced with permission from [3].

an intrinsic upper limit of $\mu_0 M_s = 2.43$ T for an $Fe_{65}Co_{35}$ alloy, where μ_0 is the permeability of free space and M_s is the saturation magnetization.

Lately, there is a much-energized interest in various types of high-performance permanent magnets based on rare-earth intermetallic compounds. This is led by, for instance, the rising need for energy-efficient technologies in which these magnets often play a vital role. The need for enlarged energy densities at different operating temperatures is the main motive for the development of the rare-earth permanent magnets (RPMs). Most importantly, this comprises less Dy-containing Nd_2Fe_{14} B-type magnets with much improved temperature stability for electromotor applications at around 450 K [4], Pr_2Fe_{14} B-type magnets for applications at 77 K together with high-Curie temperature (Tc) superconductors, [5] and a new generation of SmCo 2:17-type magnets which are applied at temperatures above 670 K [6, 7]. It also includes magnetic-power microelectromechanical systems (MEMSs) [8–11], for example, a high-speed permanent magnetic generator that requires textured, thick RPM films [12]. Currently, importance of research is on how to control the structure of grain boundary phases to understand the relevant coercivity mechanisms and the related elementary magnetization processes. The next class of permanent magnets could be rough-surfaced nanocomposites. This would include controlling the fabrication of privately mixed multiphase and well-directed nanoscale magnets, which cannot be done by conventional techniques.

1.1.2 Soft magnetic materials

The most characterizing properties of soft magnetic materials are the easy magnetization reversal accompanied with a small area of the hysteresis loop and a low coercivity (He). Quite similar to hard magnetic materials, essential magnetic properties and microstructure are to be optimized to obtain soft magnetic materials. However, a very low magnetocrystalline anisotropy and weak to almost zero interaction between magnetic domain walls and grain boundaries are required, which is the opposite of the favorable conditions for permanent magnets. Soft magnetic materials are very significant for the subjects of power electrical applications such as generators, distribution transformers, and a broad assortment of motors as well as in electronics where a mass of inductive components is required as shown in the road map of ultra-low-loss nanocrystalline alloy as shown in **Figure 3** [13]. The widely used soft magnetic materials are low-carbon steel and non-oriented silicon iron. They account for about 80% by weight, and approximately 55% by value of all soft magnetic materials, followed by grain/oriented silicon iron (17/13%), ferrite cores (1.5/7.5%), nickel- and cobalt-iron alloys (0.5/4.5%), and special materials and offices such as metal powder cores (2/8%). Soft magnetic materials are materials easily magnetized and demagnetized. They typically have intrinsic coercivity less than 1000 A m^{-1} and they are used to enhance and/or channel the flux created by an electric current. The main parameter for soft magnetic materials is the relative permeability (μr, where $\mu_r = B/\mu_0 H$), which measures the material response to the applied magnetic field. The other important parameters are the coercivity, the saturation magnetization, and the electrical conductivity. The applications for soft magnetic materials are divided into two main categories: AC and DC. In DC applications, the material is magnetized in order to carry out an operation and then demagnetized at the end of the operation, for example, an electromagnet on a lift at a scrap yard will be switched on to attract the scrap steel and then switched off to drop the steel.

For DC applications, the main regard for material selection is very likely to be the permeability. Where the material is used to produce a magnetic field or to create a force, the saturation magnetization may also be important. For AC applications, the important thought is how much energy is lost in the system as the material is cycled around its hysteresis loop. The energy loss can arise from three different sources: (1) hysteresis loss, which is related to the area contained within the hysteresis loop; (2) eddy current loss, related to the generation of electric currents in the magnetic material and the interrelated resistive losses; and (3) irregular loss, related to the movement of domain walls within the material.

Soft magnetic alloys have competed a key role in power generation and conversion for the electrical grid. The necessity for efficient generation, transmission, and distribution of electric power is ever growing; but, at the same time, the annual electric losses are overtaking annual increases in electricity consumption. In the USA, electricity is regenerated to high-voltage AC current at voltages between 138 and 765 kV and transmitted to substations close to its end-use location. The voltage is then turned down to lower values (between 13 kV and 120 V) for distribution to

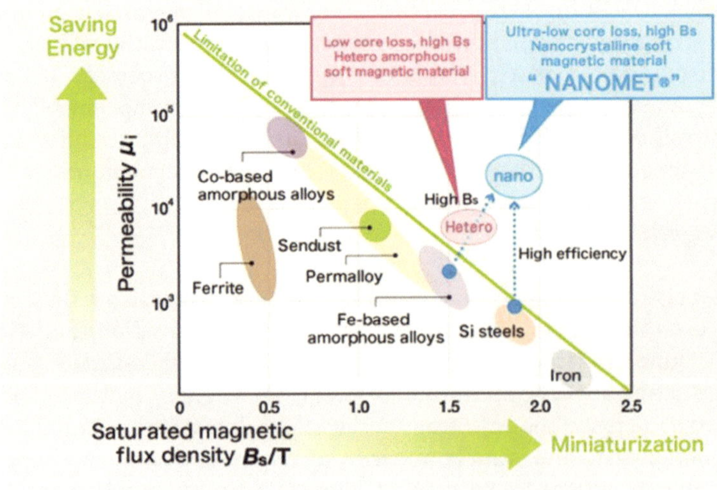

Figure 3.
Development road map of ultralow-loss nanocrystalline alloy. Reproduced with permission from [13].

different consumers. These generation, transmission, and distribution systems are aging, inept, and imperfect to meet the future energy needs of the USA without important changes in operation and infrastructure. For these reasons, advanced electric storage systems, smart controls, and power electronics for AC-DC conversion are technologies that are being supported to reform the desired way.

2. Magnetocaloric materials

Modern society depends on readily available refrigeration for preserving food and providing comfortable living places. Ordinary refrigerators use ozone for reducing harmful chemicals such as chlorofluorocarbons (CFCs), hydrofluorocarbons (HFCs), and ammonia (NH_3) in a vapor compression cycle to supply cooling. Ordinary refrigerators tend to be unwieldy, hefty, and lack energy efficiency despite they have met the cooling needs. Recently, an alternate refrigeration method using magnetocaloric effect (MCE) has been investigated as a way to deal with the defects of vapor-compression refrigeration.

Magnetic refrigeration has three outstanding advantages when compared to gas compressing refrigeration. First, it involves no harmful gasses; second, it can be compactly built as its main working material is a solid; and third, magnetic refrigerators are almost noiseless. Also, the cooling efficiency while operating with gadolinium can reach 60% of the theoretical efficiency limit [14] compared to only about 45% in the best gas-compressing refrigerators. While commercial refrigerators of this kind are still in the development stages, research efforts to develop new materials with improved MCE are targeted on maximizing the cooling capability and energy efficiency of this newborn technology. In this part, the different materials are compared, focusing on transition metal-containing compounds. When a material is subjected to an applied magnetic field, its magnetic order changes, leading to subsequent change of the entropy related to the magnetic degrees of freedom (magnetic entropy, S_m). Under adiabatic conditions, ΔS_m must be covered by an equal, opposite change in the entropy associated with the lattice, resulting in a change in the temperature of the material. This temperature change, ΔT_{ad}, is usually called the MCE. It is correlated to the magnetic properties of the material through the thermodynamic Maxwell relation

$$\left(\frac{\partial s}{\partial B}\right)_T = \left(\frac{\partial M}{\partial T}\right)_B \tag{1}$$

From magnetization measurements taken at different temperature periods, ΔS_m can be calculated as illustrated in Refs. [15, 16]. For materials showing a first-order phase transition with large hysteresis, these magnetization measurements should be performed cautiously so as not to overestimate values of the entropy change [17]. Otherwise, the magnetic entropy change can be acquired straight from a calorimetric measurement of the field dependence of the high temperature capacity, c, and then integrating. It has been validated that the values of ΔS_m (T, B) derived from the magnetization measurement concur with the values from calorimetric measurement [18]. Numerical integration of the adiabatic temperature change, [ΔT_{ad} (T, B)], can then be done using the experimentally or theoretically predicted magnetization and heat content values. Clearly, the MCE will be large when $\left(\frac{\partial M}{\partial T}\right)_B$ is large and c (T,B) is small at the same temperature conditions. As effects at high temperatures are concerned, the heat capacity on the order of Dulong-Petit law is c = 3 NR, where N is the number of atoms and R is the molar gas constant. Consequently, we should focus on finding a big change in magnetization at the

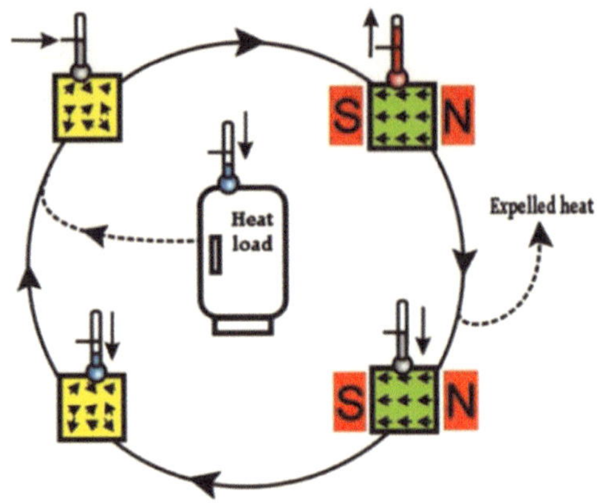

Figure 4.
Schematic representation of a magnetic refrigeration cycle that transports heat from the heat load to the ambient environment. Yellow and green boxes depict materials in low and high magnetic fields, respectively. Reproduced with permission [20]. Copyright 2005, Institute of Physics.

appropriate temperature. A large MCE is anticipated not far from $\left(\frac{\partial M}{\partial T}\right)_B$ peaks at the magnetic-ordering temperature since the order parameter of the phase transition changes intensely within a narrow temperature interval. In the magnetic-refrigeration cycle, shown in **Figure 4** [19, 20], initial random-oriented magnetic moments are ordered by a magnetic field, resulting in heating of the magnetocaloric material and the heat is then transmitted from the material to the surrounding atmosphere. Upon removing the field, the magnetic moments disorder resulting in cooling of the material below ambient temperature. Heat from the system can then be withdrawn by a heat-transfer medium which may be water, air, or helium depending on the working temperature. Consequently, magnetic refrigeration is considered an ecofriendly cooling technology.

3. Magnetic nanoparticles

Over time, nanotechnology has penetrated all branches of science like physics, chemistry, and especially biomedical research and related industries. Broadly, nanoparticles are defined as materials having particle sizes in the range of 1–100 nm [21]. Bulk materials have definite physical properties, which, however, get altered when they are converted to nanoparticles, depending on their final size. One of the main changes in the properties of nanoparticles is the substantial increase in number of atoms/molecules on the surface of particles, and hence availability of effectively high surface area compared with bulk material. The high surface area of particles can be used to attach ligands and/or capping agents, which make them more suitable for effective labeling of drug/tracer molecules. The change in physicochemical properties during conversion of bulk material to nanoparticles makes them suitable for reaching the diseased site because of their better diffusion ability. A diversity of nanoparticles, including magnetic nanoparticles (MNs), has been synthesized and characterized for different industrial, biomedical, and clinical applications.

MNs are the nanoparticles synthesized from magnetic elements like iron, nickel, and cobalt or their chemical derivatives [21–26]. Each particle of bulk magnetic materials has many domains separated by walls, and each domain represents a region with a specific direction of magnetization. When bulk material is converted to MN, each particle can approach a single domain [22–24]. In larger particles (micrometer size), surrounding thermal energy [kT, where k is the Boltzmann constant and T is the temperature (K)] is much less [when T = 300 K (room temperature), kT = 0.026 eV] than particle energy (Kv, where K is the anisotropic constant and v is the particle volume) and thus the direction of magnetic moment does not change with time. When particle size decreases (sub-micro-meter size), particle energy decreases and thus direction of magnetic moment also changes with respect to original direction, that is, with angle (θ). However, with further decrease of particle size (nanosize), the direction of magnetic moment changes to the opposite direction (θ = 180), which is known as superparamagnetic behavior of magnetic nanoparticles. Super paramagnetism is due to particle size, whereas paramagnetism is an intrinsic property of the material caused by its atomic nature (e.g., Na). Superparamagnetic particles have high magnetic moment of 10^3–10^4 μ_B [27, 28] and thus the term "super" is prefixed to "paramagnetic" because particles show paramagnetic behavior in the absence of a magnetic field and no magnetization is retained after removal of the magnetic field. Decreasing particle size below the critical size, ferromagnetic particles can be changed to superparamagnetic particles. Paramagnetic materials (e.g., Na and K) [22] do not have magnetic interactions between the atoms; hence, the net magnetic moment is equivalent to the number of atoms in the particle. However, the interatomic magnetic interaction in ferromagnetic or superparamagnetic materials gives the net magnetic moment of the particle. On either decreasing temperature or increasing magnetic field, there is a possibility of transition from superparamagnetic to ferromagnetic (**Figure 5**) [29, 30] because of increasing extent of the arrangement of spins of MN.

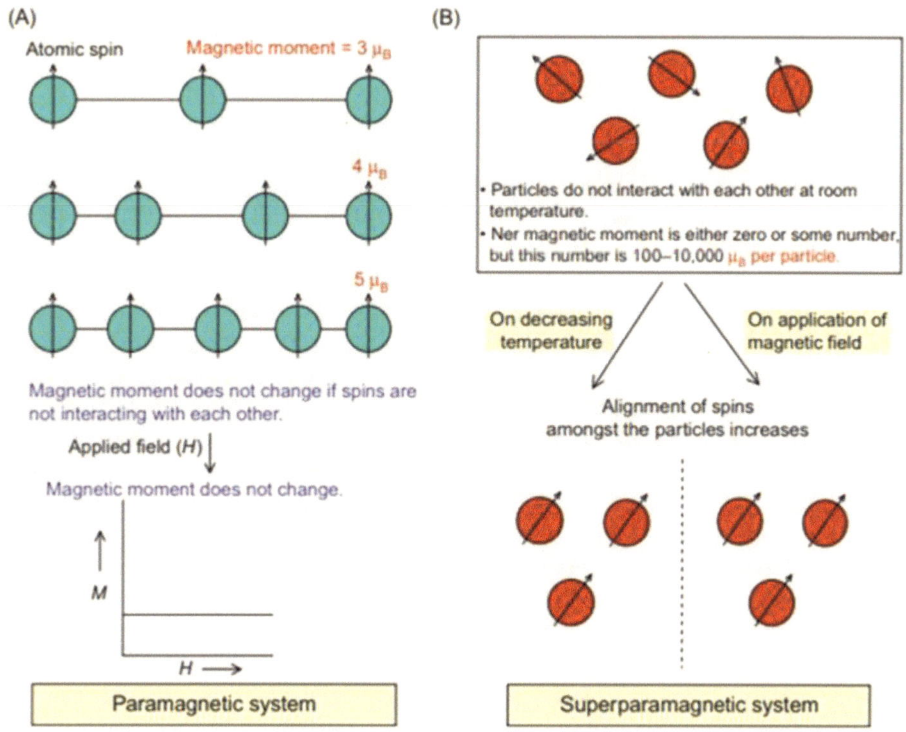

Figure 5.
(A) Paramagnetic particles under a magnetic field. No variation of magnetization is shown and (B) superparamagnetic particles under a magnetic field or at low temperature [30]. Copyright 2003, Institute of Physics.

Owing to their unique feature of attraction and interaction under magnetic field conditions, these MNs have been applied for separation of cells/biological materials and drug delivery. MNs have attracted the researchers' attention because of their ability to act as contrast agents in magnetic resonance imaging (MRI) for diagnostic applications. It may be apposite to observe here that lower toxicity, biocompatibility, and significant accumulations of MNs at the diseased site make them suited for remedial applications. When these MNs are placed under magnetic field effects, a phase interval between the applied magnetic field and the direction of magnetic moments results in thermal losses. The orientation of magnetic moment fluctuates thermally, involving two main mechanisms: (i) Neel's fluctuations of the magnetic moment relative to the crystal lattice (internal dynamics) and (ii) Brownian fluctuations of the particle itself relative to the medium in which the particle is placed (external dynamics). These are affected by viscosity of the medium and other processes, which can affect the movement of particle. These external and internal frictions generated on MN under external magnetic field conditions result in "foci" of heat generation, which may be sufficient enough to kill the cell. Thus, selective heat generation by MN at the tumor site can provide the significant advantage of killing tumor cells without affecting the normal tissues much.

4. Applications of functional magnetic nanoparticles

The unique chance to control coercivity in magnetic nanomaterials has led to a number of significant technological applications, particularly in the field of information storage. Small magnetic particles are promising candidates for a further increase of the density of magnetic storage devices toward 100 $G_{bit}/inch^2$ up to a few $T_{bit}/inch^2$ [31]. Other than data storage, many applications of magnetic nanoparticles are known; examples are: ferrofluids, high-frequency electronics, high-performance permanent magnets, and magnetic refrigeration. Magnetic particles are also employed in many biological and medical applications such as drug-targeting, cancer therapy, lymph node imaging, or hyperthermia [32–34]. Lately, researchers have succeeded to produce multifunctional MN. There are mainly two approaches: (i) molecular functionalization, which comprises attaching the magnetic nanoparticles to antibodies, proteins, and dyes, and so on and (ii) blending of MNs with other functional nanoparticles, such as quantum dots or metallic nanoparticles [35]. As an example, magnetic nanoparticles could be used as seeds for growing semiconducting chalcogenides. In this case, the final product is core-shell or hetero nanostructures having both magnetic and fluorescent properties. This results in the display of intracellular control of nanoparticles for promising dual-functional molecular imaging (i.e., combined MRI and fluorescence imaging). MNs can be used as MRI contrast improvement agents, as the signal resulting from proton magnetic moments around magnetic nanoparticles can be recorded by resonant absorption [24]. These multifunctional MNs could be used in many biological applications such as protein purification, bacteria detection, and therapeutic removal of toxins [32]. **Figure 6** illustrates these two approaches for making multifunctional MNs and their various biological applications.

In the last three decades, magnetic data storage has seen a linear rise in terms of storage capacity. The physics of magnetic nanostructures is at the heart of magnetic hard disk drive technology. In the future, it is very probable that areal densities will increase well beyond 1 Terabit/inch^2 by employing new technologies like bit-patterned media (BPM) or heat-assisted magnetic recording [31, 36].

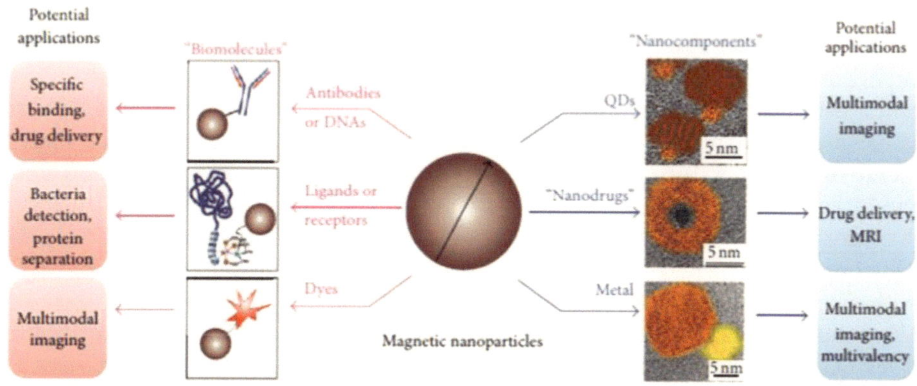

Figure 6.
Various potential applications of multifunctional magnetic nanoparticles in biology. Reproduced with permission from [24]. Copyright 2009, American Chemical Society.

Patterned magnetic nanostructures, such as two-dimensional dot-arrays have attracted the interest of researchers due to their potential applications such as magnetic information storage [37] or nonvolatile magnetic random access memory (MRAM) [38]. The demand for ultrahigh-density magnetic storage devices drives the bit size into the nanometer scale. As the volume $V = \pi D2t/4$ (where D and t are the diameter and thickness, respectively) of the grains is reduced in the scaling process, the magnetization of the grains may become unstable due to thermal fluctuations, and data loss may occur [33]. As the physical size of the nanostructures in the patterned array decreases, loss of data due to the thermal instability [also known as "superparamagnetic (SPM) effect"] would become a very crucial issue [39]. Therefore, future data storage technology has to overcome the SPM effect. In this regard, the L10-FePt alloy is one of the most promising materials for future ultrahigh-density magnetic storage devices because it possesses a huge uniaxial magneto-crystalline anisotropy ($Ku = 7 \times 10^7$ erg/cc), which leads to a high thermal stability of magnetization. Also, the present longitudinal data storage media may be considered as a collection of independent particles because of their weak intergranular exchange coupling. However, as we have discussed in the super-ferromagnetic section, strong intergranular interactions can drive the system to form long-range ordered super-ferromagnetic (SFM) domains, which are clearly unsuitable for applications in data storage. Also, the SFM alignment counteracts large tunneling magnetoresistance (TMR) values, so magnetic random access memory applications are not promising for SFM systems. However, super-ferromagnetic materials are soft magnetics, which make them nearly ideal materials for high permeability, low-loss materials for micro-electronics, power management, and sensing devices designed for high frequencies.

4.1 Magnetic materials in hyperthermia

Recently, thermotherapy for cancer using MN has emerged as a potential mode of hyperthermia [23–26]. Hyperthermia is a type of medical treatment in which body tissue is exposed to a temperature (42–44°C) higher than physiological temperature (37°C) to kill the cancer cells. This approach is one of the modalities of cancer treatment used in combination with radiation and certain chemotherapeutic drugs. There could be two ways to heat the cancer cells: (i) applying external sources (e.g., using a water bath, microwave, ultrasound, infrared sauna), which is also called "external or extracellular hyperthermia," and (ii) delivering MN inside the cancer cells [under alternating current (AC) field], which is known as intracellular hyperthermia. Because cell membrane composed of lipids is thermally insulating, tumor cells heated from external sources do not achieve hyperthermic temperature. Consequently, extra heat from an external source has to be provided to achieve the therapeutic temperature. However, this causes blisters, burns, swelling, blood clots, and bleeding in clinical conditions. Therefore, application of hyperthermia using this approach has faced practical limitations. On the other hand, intracellular heating using internalized MN at the tumor site provides an efficient and safe approach for hyperthermia application. The therapeutic efficacy and clinical advantages of intracellular hyperthermia over extracellular hyperthermia is a matter of further investigation. In addition, development of surface-functionalized nanoparticles using advanced technologies may present a better therapeutic modality for future clinical applications. Could all MNs be used in hyperthermia? Common MNs are Fe_3O_4; γ-Fe_2O_3; and Mn-, Co-, and Ni-doped ferrites because they have high magnetic moment (50–60 emu/g) under external magnetic field, which can give hysteresis loss and result in significant rise in temperature sufficient for hyperthermia therapy. However, some materials (e.g., ZnO and TiO_2) become ferromagnetic when particle size decreases to the nanometer range (510 nm) [40, 41]. Owing to their very low magnetic moment (1 emu/g or less), such types of material may not be useful for hyperthermia treatment. It may be important to mention that Fe and Co nanoparticles are prone to oxidation in acidic and alkaline conditions, which are likely to be different in tissue compartments in body. In contrast, oxide nanoparticles (e.g., Fe_3O_4) are highly stable in slightly acidic and alkaline conditions and are biocompatible. Very small Fe_3O_4 (cubic phase) nanoparticles (5 nm) are not useful for hyperthermic applications because of low magnetic moment [29, 30]. However, FePd, FePt, CoPt, and CoPd (tetragonal phase) nanoparticles would result in significant heat generation, even with a particle size of 35 nm [27], but their stabilities in acidic and alkaline mediums are less than their oxide counterparts.

4.2 Magnetic materials in data storage

Magnetic materials are used in high-capacity disk drives and magnetic-semiconductor memory devices. The disk drive devices have reached the largest growth in data capacity over time, making disk drives the preeminent storage system for digital data [42]. The growth in areal density is more than 100% per year recently. The overall data capacity of a disk is nearly the areal density times the recording area depending on the disk size (the most common diameter is 2.5 and 3.5 inches, that is, 64 and 90 mm, respectively). Many technologies have aided in this speedy increase in areal density, together with enhancement of the technology of "flying" heads with shrunk space of the disk surface, data coding, error discovery and rectification, advanced servo-control systems for correct management of magnetic recording heads on data tracks, and advances in the mechanical structures comprising a disk drive, together with advances in the motors used to push the disks. Recently, there has been a significant emerging technology for fast memory devices—the magnetic random-access memory or MRAM. The MRAM device is a possible substitute for the familiar semiconductor memories used in modern computers—dynamic and static random-access memory (DRAM and SRAM). The MRAM technology combines a magnetic storage technology together with metal-oxide semiconductor (MOS) devices to result in fast and high-density data memory devices. The technology on which the magnetic part of MRAM is based is an extension of the technology used in magnetic-recording devices identified as the magnetic tunneling junction or MTJ.

5. Technology of magnetic recording

The technology of magnetic recording is over one century old [43]. The fundamental concept of magnetic recording is to use a magnetic structure (as the "write" head) driven by a current that represents the data to be recorded, to create a magnetic field capable of changing the state of the magnetization in a closely spaced magnetic medium, which was formerly a magnetic wire, and today it is the known tape or a magnetic layered hard drive. The data are retrieved by an output electromotive force generated in the "red head" by sensing the magnetization in the recording medium, for example, by Faraday's law. The magnetic recording system is that used to store digital data, in which instance the current supplied to the write head as pulses coded to represent the digital information (1 or 0 s) [44–47]. In the case of disk drives, the write and read heads are distinct thin-film structures deposited on the back of a mechanical slider that uses a hydrodynamic air bearing to "fly" over the surface of the disk [46]. The read and write parts are viewed together with the magnetic recording surface, which is a thin cobalt metal alloy film. The digital data are written in the magnetic film in the form of transitions among the two magnetization states (the "left" or "right") and with the width almost equal to the write head width. The transition region between the oppositely directed directions of the magnetization is similar to that between magnetic domains and has a length (l). The write head is formed from thin films of ferromagnetic alloys patterned in the form of a magnetic chain. The current is coupled to the chain to generate a magnetic field at the gap by a pancake coil of 10 or less turns. The coil is insulated from the metallic magnetic bondage by layers of polymer photoresist. The most frequently used alloy in the past for the magnetic films in the write head is $Ni_{80}Fe_{20}$ permalloy which can be deposited in thin films using electroplating. The ability to record on recording media with increased coercivity is not the only issue with the magnetic materials used in write heads. It is also important that the write head have high efficiency. Efficiency (η) in this case is defined as the ratio

$$\eta = \frac{H_g \cdot g}{N_\omega \cdot I} \tag{2}$$

where H_g is the value of the magnetic field in the gap of the write head and I is the amplitude of the write current pulse. High efficiency is important to allow write-current amplitudes that are easily supplied from integrated circuits.

6. Summary and perspectives

Functional magnetic materials are a huge source of technological applications because they can simultaneously display intriguing properties such as tunable mechanical, magnetic, electric/dielectric, thermal, and optical properties. These materials have the potential to be used in information storage and processing, refrigeration, hyperthermia, and recording technology. Though most attention is paid to the pure magnetocaloric properties and materials costs, other properties like mechanical properties, heat conductivity, electrical resistivity, and environmental impact are recently getting attention. With the refrigeration market being a multibillion dollar market, this novel technology offers great opportunities. The ideal magnetic refrigerant should contain at least 80% transition metals having large magnetic moment such as Fe or Mn. In addition, it should contain some inexpensive p-metal such as Al or Si, which can be used to tune the working point of the material. A wide range of magnetic materials is essential for the advance of magnetic recording and the fast random access memory, MRAM, technology. Magnetic data storage has seen a linear rise in terms of storage capacity. The physics of magnetic nanostructures is at the core of magnetic hard disk drive technology; and in the future, it is very likely that areal densities will increase well beyond 1 Terabit/inch^2 by employing new technologies. In hyperthermia application, the target is the higher value of magnetic heat generation by a stable fluid in a lower exposure time. Nanoferrites are good candidates for hyperthermia applications since they offer a moderate magnetic moment, chemical stability, and a high specific absorption rate (SAR). Based on which heat generation mechanism is wanted, a suitable selection of magnetic core, surfactant layer, and liquid type can influence the cancer treatment.

Acknowledgements

The authors would like to extend their sincere appreciation to Central Metallurgical Research and Development Institute, Egypt, for its financial support to pursue this work.

Author details

Sabah M. Abdelbasir* and Ahmed Esmail Shalan
Central Metallurgical Research and Development Institute (CMRDI), Helwan, Cairo, Egypt

*Address all correspondence to: sfoda20@hotmail.com

References

[1] Satoshi H, Masamichi N, Seiji M. Perspectives for high-performance permanent magnets: Applications, coercivity, and new materials. Advances in Natural Sciences: Nanoscience and Nanotechnology. 2017;**8**:013002. DOI: 10.1088/2043-6254/aa597c

[2] Gutfleisch O. Controlling the properties of high energy density permanent magnetic materials by different processing routes. Journal of Physics D: Applied Physics. 2000;**33**:R157

[3] http://power.eee.metu.edu.tr.

[4] Sugimoto S. In: Niarchos D, editor. Proc. 20th Int. Workshop on Rare Earth Permanent Magnets and their Applications. 2008. p. 106

[5] Hinz D, Lyubina J, Fuchs G, Brown DN, Ma BM, Gutfleisch O, et al. Hot deformed (Nd,Pr)(Fe,Co)B magnets for low temperature applications. Journal of Magnetism and Magnetic Materials. 2004;**272-276**:e321-e322

[6] Hai-Peng Chen C, Lee D, Liu S, Walmer M, Zhang Y, Hadjipanayis GC. Fully dense bulk nanocomposites of $(Sm_{1-x}Gd_x)_2(Co_{1-y}Fe_y)_{17}$+(Co, Fe) with high coercivity and high Curie temperature. IEEE Transactions on Magnetics. 2004;**40**:2937

[7] Gutfleisch O, Müller KH, Khlopkov K, Wolf M, Yan A, Schäfer R, et al. Evolution of magnetic domain structures and coercivity in high-performance Sm Co 2:17-type permanent magnets. Acta Materialia. 2006;**54**:997

[8] Holmes AS, Hong G, Pullen KR. Axial-flux permanent magnet machines for micropower generation. Journal of Microelectromechanical Systems. 2005;**14**:54-62

[9] Raisigel H, Cugat O, Delamare J. Permanent magnet planar micro-generators. Sensors and Actuators, A: Physical. 2006;**103-131**:438-444

[10] Das S, Arnold DP, Zana I, Park JW, Lang JH, Allen MG. Multi-watt electric power from a microfabricated permanent magnet generator. In: IEEE MEMS 2005 Conference; Miami, USA. 2005. pp. 287-290

[11] Gutfleisch O, Dempsey N. In: Azzerboni B, Asti G, Pareti L, Ghidini M, editors. Magnetic Nanostructures in Modern Technology: Spintronics. Berlin: Springer; 2008. p. 167

[12] Dempsey NM, Walther A, May F, Givord D, Khlopkov K, Gutfleisch O. High performance hard magnetic NdFeB thick films for integration into micro-electro-mechanical systems. Applied Physics Letters. 2007;**90**:092509

[13] Available from: http://nanoc.imr.tohoku.ac.jp/eng/research.html

[14] Zimm C, Jastrab A, Sternberg A, Pecharsky V, Geschneidner K Jr, Anderson MI. Description and performance of a near-room temperature magnetic refrigerator. Advances in Cryogenic Engineering. 1998;**43**:1759-1766

[15] Tishin AM, Spichkin YI. The Magnetocaloric Effect and Its Applications, Institute of Physics. Bristol: IOP Publishing Ltd; 2003

[16] Brück E. In: Buschow KHJ, editor. Handbook of Magnetic Materials. Vol. 17. North Holland, Amsterdam: Elsevier; 2008. pp. 235-291

[17] Caron L, Ou ZQ, Nguyen TT, Thanh DTC, Tegus O, Bruck E. On the determination of the magnetic entropy change in materials with first-order

transitions. Journal of Magnetism and Magnetic Materials. 2009;**321**: 3559-3566

[18] Gschneidner KA, Pecharsky VK, Pecharsky AO, Zimm CB. Recent developments in magnetic refrigeration. Rare Earths '98. 1999;**315-317**:69-76

[19] Tegus O, Brück E, Buschow KHJ, de Boer FR. Transition-metal-based magnetic refrigerants for room-temperature applications. Nature. 2002;**415**:150-152

[20] Brück E. Developments in magnetocaloric refrigeration. Journal of Physics D: Applied Physics. 2005;**38**:R381

[21] Gleiter H. Nanocrystalline materials. Progress in Materials Science. 1989;**33**:223-315

[22] Cullity BD, Graham CD. Introduction to Magnetic Materials, 2nd ed. Hoboken, New Jersey: John Wiley & Sons, Inc., Publication; 2009

[23] Bréchignac C, Houdy P, Lahmani M. Nanomaterials and Nanochemistry. Berlin, Heidelberg: European Materials Research Society, Springer-Verlag; 2007

[24] Miller JS, Drillon M. Magnetism: Molecules to Materials III. Weinheim (Federal Republic of Germany): Wiley-VCH Verlag GmbH; 2002

[25] Ohno T, Wakabayashi T, Takemura A, Yoshida J, Ito A, Shinkai M, et al. Effective solitary hyperthermia treatment of malignant glioma using stick type CMC-magnetite. in vivo study. Journal of Neuro-Oncology. 2002;**56**:233-239

[26] De Greef M, Kok HP, Correia D, Bel A, Crezee J. Optimization in hyperthermia treatment planning: The impact of tissue perfusion uncertainty. Medical Physics. 2010;**37**:4540-4550

[27] Gajbhiye NS, Sachil S, Ningthoujam RS. Synthesis of self-assembled monodisperse 3 nm sized FePd nanoparticles: Phase transition, magnetic study and surface effect. Journal of Applied Physics. 2008;**104**:123906-123907

[28] Ningthoujam RS, Gajbhiye NS. Magnetization studies on ε-Fe2.4Co0.6N nanoparticles. Materials Research Bulletin. 2010;**45**:499-504

[29] Goya GF, Berquo TS, Fonseca FC, Morales MP. Static and dynamic magnetic properties of spherical magnetite nanoparticles. Journal of Applied Physics. 2003;**94**:3520-3529

[30] Barbeta V, Jardim BRF, Kiyohara PK, Effenberger FB, Rossi LM. Magnetic properties of Fe_3O_4 nanoparticles coated with oleic and dodecanoic acids. Journal of Applied Physics. 2010;**107**:073913-073917

[31] Moser A, Takano K, Margulies DT, Albrecht M, Sonobe Y, Ikeda Y, et al. Magnetic recording: Advancing into the future. Journal of Physics D. 2002;**35**(19):R157-R167

[32] Berry CC, ASG C. Functionalization of magnetic nanoparticles for applications in biomedicine. Journal of Physics D. 2003;**36**(13):R198-R206

[33] Tartaj P, del Puerto Morales M, Veintemillas-Verdaguer S, Gonźalez-Carreno T, Serna CJ. The preparation of magnetic nanoparticles for applications in biomedicine. Journal of Physics D. 2003;**36**(13):R182-R197

[34] Andra W, Hafeli U, Hergt R, Misri R. Application of magnetic particles in medicine and biology. In: Kronmüller H, Parkin SPS, editors. Handbook of Magnetism and Advanced Magnetic Materials, Novel Materials. Vol. 4. Chichester, UK: John Wiley & Sons; 2007. pp. 22536-22568

[35] Euliss LE, Grancharov SG, O'Brien S, Deming TJ, Stucky GD, Murray CB, et al. Cooperative assembly of magnetic nanoparticles and block copolypeptides in aqueous media. Nano Letters. 2003;**3**(11):1489-1493

[36] Shiroishi Y, Fukuda K, Tagawa I, Iwasaki H, Takenoiri S, Tanaka H, et al. Future options for HDD storage. IEEE Transactions on Magnetics. 2009;**45**(10):3816-3822

[37] Terris BD, Thomson T. Nanofabricated and self-assembled magnetic structures as data storage media. Journal of Physics D. 2005;**38**(12):R199-R222

[38] Guo Y, Wang P, Chen M-M, Horng C, Min T, Hong L, et al. MRAM array with coupled soft-adjacent magnetic layer. Journal of Applied Physics. 2005;**97**(10):10P506

[39] Weller D, Moser A. Thermal effect limits in ultra-high density magnetic recording. IEEE Transactions on Magnetics. 1999;**35**(6):4423-4439

[40] Ningthoujam RS, Gajbhiye NS, Ahmed A, Umre SS, Sharma SJ. Re-dispersible Li^+ and Eu^{3+} co-doped nanocrystalline ZnO: Luminescence and EPR studies. Journal of Nanoscience and Nanotechnology. 2008;**8**:3059-3062

[41] Panigrahy B, Aslam M, Misra DS, Ghosh M, Bahadur D. Defect-related emissions and magnetization properties of ZnO nanorods. Advanced Functional Materials. 2010;**20**:1161-1165

[42] Grochwski E, Almaden IBM. Research Laboratories. Available from: http://www.storage.ibm.com/technolo/grochows/grocho01.htm

[43] Mee CD, Daniel ED, Clark MH. Magnetic Recording: The First 100 Years. New York: John Wiley & Sons; 1998

[44] Comstock RL. Introduction to Magnetism and Magnetic Recording. New York: John Wiley & Sons; 1999

[45] Comstock RL. Data storage in rigid disks. In: Mee CD, Daniel ED, editors. Magnetic Storage Handbook. 2nd ed. New York: McGraw-Hill; 1996

[46] Mee CD, Daniels ED. Magnetic Recording Technology. New York: McGraw-Hill; 1996

[47] Wang SX, Taratorin AM. Magnetic Information Storage Technology. San Diego: Academic Press; 1999

Chapter 5

Control of Spin-Orbit Interaction in Conventional Semiconductor Quantum Wells

Jiyong Fu, Wei Wang and Minghua Zhang

Abstract

In this chapter, we demonstrate, focusing on GaAs quantum wells (QWs), a full control of spin-orbit (SO) interaction including both the Rashba and Dresselhaus terms in conventional semiconductor QWs. We determine the SO interaction in GaAs from single to double and triple wells, involving the electron occupation of either one or two subbands. Both the intraband and interband SO coefficients are computed. Two distinct regimes, depending on the QW width, for the control of SO terms, are found. Furthermore, we determine the persistent-spin-helix (PSH) symmetry points, where the Rashba and the renormalized (due to cubic corrections) Dresselhaus couplings are matched. These PSH symmetry points, at which quantum transport is diffusive (2D) for charge while ballistic (1D) for spin, are important for longtime and long-distance coherent spin control that is the keystone in spintronic devices.

Keywords: spintronics, spin-orbit interaction, Rashba term, Dresselhaus term, persistent spin helix, semiconductor, quantum well

1. Introduction

The spin-orbit (SO) interaction is a relativistic effect coupling spatial and spin degree of freedom via an effective magnetic field, facilitating spin manipulation in semiconductor nanostructures [1, 2]. For instance, the proposal of Datta and Das for a spin field-effect transistor highlights the use of the SO interaction of Rashba [3]. Recently, the SO effects [4] have attracted renewed interest in diverse fields of condensed matter, including the persistent spin helix (PSH) [5–8], topological insulators [9], and Majorana fermions [10, 11].

In zinc-blende-type crystals, such as GaAs, there are two dominant contributions to the SO interaction. The bulk inversion asymmetry leads to the Dresselhaus coupling [12], which in heterostructures contains both linear and cubic terms. The linear term mainly depends on the quantum-well confinement and the cubic one on the electron density [7, 13]. Additionally, the structural inversion asymmetry in heterostructures gives rise to the linear Rashba coupling [14], which can be electrically controlled by using an external bias [15, 16]. Extensive studies on the SO interaction have been focused on n-type GaAs/AlGaAs wells with only one-subband electron occupation [7, 13, 17]. Recently, quantum wells with two populated subbands have also drawn attention in both experiment [18–20] and theory [21–25], because of emerging new physical phenomena including the intersubband coupling-induced spin mixing [19] and *crossed* spin helices [25].

In this chapter, we report our recent results on the electric control of SO interactions in conventional semiconductor quantum wells. *Firstly*, we focus on the case of single GaAs wells and performed a detailed self-consistent calculation to determine how the SO coupling (both the magnitude and sign) changes as a function of the gate voltage V_g. The gate-altered electron occupations between the first two subbands have been taken into account, and both the intraband and interband SO coefficients are computed. We find two distinct regimes for the electric control of SO interactions. *Secondly*, we consider the case of multiple wells and determine the SO interaction in GaAs from single to double and triple wells. Furthermore, we determine the persistent-spin-helix (PSH) symmetry points, where the Rashba and the renormalized (due to cubic corrections) Dresselhaus couplings are matched. These PSH symmetry points, at which quantum transport is diffusive (2D) for charge while ballistic (1D) for spin, are important for longtime and long-distance coherent spin control that is the keystone in spintronic devices.

2. Model Hamiltonian

The quantum wells that we consider are grown along the $z\|[001]$ direction. We start from the 8×8 Kane model, involving both conduction and valence bands, and use the folding down procedure [4, 24], to obtain an effective 3D Hamiltonian for electrons only [4, 13, 24]:

$$\mathcal{H}^{3D} = \frac{\hbar^2 k_\|^2}{2m^*} - \frac{\hbar^2}{2m^*}\frac{\partial^2}{\partial z^2} + V_{sc}(z) + \mathcal{H}_\mathcal{R} + \mathcal{H}_\mathcal{D}, \tag{1}$$

where the first two terms refer to the kinetic contributions, in which m^* is the electron effective mass and $k_\| = \sqrt{k_x^2 + k_y^2}$ is the in-plane electron momentum with $x\|[100]$ and $y\|[010]$ directions. The third term V_{sc} describes the confining electron potential with the subscript sc indicating that it is determined self-consistently within the (Poisson-Schrödinger) Hartree approximation. The confining potential contains the structural potential V_w arising from the band offsets, the doping potential V_d, the electron Hartree potential V_e, and the external gate potential V_g [13, 23, 24]. The last two terms, $\mathcal{H}_\mathcal{R}$ and $\mathcal{H}_\mathcal{D}$, correspond to the Rashba and Dresselhaus SO interactions, respectively. Note that in Eq. (1), we have added the Dresselhaus term $\mathcal{H}_\mathcal{D}$ empirically [4, 13, 24], since to obtain $\mathcal{H}_\mathcal{D}$ one has to take into account the remote bands (e.g., **p**-conduction band) [4]. The Rashba term reads $\mathcal{H}_\mathcal{R} = \eta(z)(k_x\sigma_y - k_y\sigma_x)$ with $\eta(z) = \eta_\mathrm{w}\partial_z V_\mathrm{w} + \eta_\mathrm{H}\partial_z(V_\mathrm{g} + V_\mathrm{d} + V_\mathrm{e})$ determining the Rashba strength and $\sigma_{x,y,z}$ the spin Pauli matrices. The constants η_w and η_H are determined by the bulk band parameters [13, 24]:

$$\eta_\mathrm{w} = \frac{P^2}{3}\left(\frac{\delta_\mathrm{v}/\delta_\mathrm{c}}{E_g^2} - \frac{\delta_\Delta/\delta_\mathrm{c}}{(E_g+\Delta)^2}\right), \tag{2}$$

$$\eta_\mathrm{H} = -\frac{P^2}{3}\left(\frac{1}{E_g^2} - \frac{1}{(E_g+\Delta)^2}\right), \tag{3}$$

where E_g is the fundamental bandgap, P is the Kane parameter, and Δ is the split-off gap, in the well layer. The parameters δ_c, δ_v, and δ_Δ stand for band offsets between well and barrier layers, of the conduction band, heavy hole (and light hole), and split-off hole, respectively. A schematic of the band offsets for a GaAs/AlGaAs well is shown in **Figure 1**. The Dresselhaus term is $\mathcal{H}_D = \gamma\left[\sigma_x k_x\left(k_y^2 - k_z^2\right) + \text{c.c.}\right]$, with γ the bulk Dresselhaus parameter and $k_z = -i\partial_z$. From the 3D Hamiltonian (Eq. (1)), we are now ready to derive an effective 2D Hamiltonian for electrons. This 2D model is similar to the well-known Rashba model but now for wells with two subbands. We first self-consistently determine the spin-degenerate eigenvalues $\mathcal{E}_{\mathbf{k}_\parallel \nu} = \mathscr{E}_\nu + \hbar^2 k_\parallel^2/2m^*$ and the corresponding eigenspinors $|\mathbf{k}_\parallel \nu\sigma\rangle = |\mathbf{k}_\parallel \nu\rangle \otimes |\sigma\rangle$, $\langle \mathbf{r}|\mathbf{k}_\parallel \nu\rangle = \exp\left(i\mathbf{k}_\parallel \cdot \mathbf{r}_\parallel\right)\psi_\nu(z)$, of the well *in the absence* of SO interaction [24]. Here we have defined $\mathcal{E}_\nu$ (ψ_ν), $\nu = 1, 2$, as the ν_{th} quantized energy level (wave function) and $\sigma = \uparrow, \downarrow$ as the electron spin component along the z direction. We then can straightforwardly obtain an effective 2D model by projecting Eq. (1) onto the basis $\{|\mathbf{k}_\parallel \nu\sigma\rangle\}$. The effective 2D model with two subbands in the coordinate system $x_+ \| (110)$, $x_- \| (1\bar{1}0)$ under the basis set $\{|\mathbf{k}_\parallel 1\uparrow\rangle, |\mathbf{k}_\parallel 1\downarrow\rangle, |\mathbf{k}_\parallel 2\uparrow\rangle, |\mathbf{k}_\parallel 2\downarrow\rangle\}$ reads [26]:

$$\mathcal{H}^{2D} = \left(\frac{\hbar^2 k_\parallel^2}{2m^*} + \mathcal{E}_+\right)\mathbf{1}\otimes\mathbf{1} - \mathcal{E}_- \tau_z \otimes \mathbf{1} + \mathcal{H}_{RD}, \tag{4}$$

with $\mathcal{E}_\pm = (\mathcal{E}_2 \pm \mathcal{E}_1)/2$, $\mathbf{1}$ the 2×2 identity matrix (in both spin and orbital subspaces), and $\tau_{x_+, x_-, z}$ the Pauli ("pseudospin") matrices acting within the orbital subspace. The term $\mathcal{H}_{RD}$ describes the Rashba and Dresselhaus SO contributions in terms of intra- and intersubband SO fields $\mathbf{B}_{SO}^{\nu}$ and $\mathbf{B}_{SO}^{12}$, respectively:

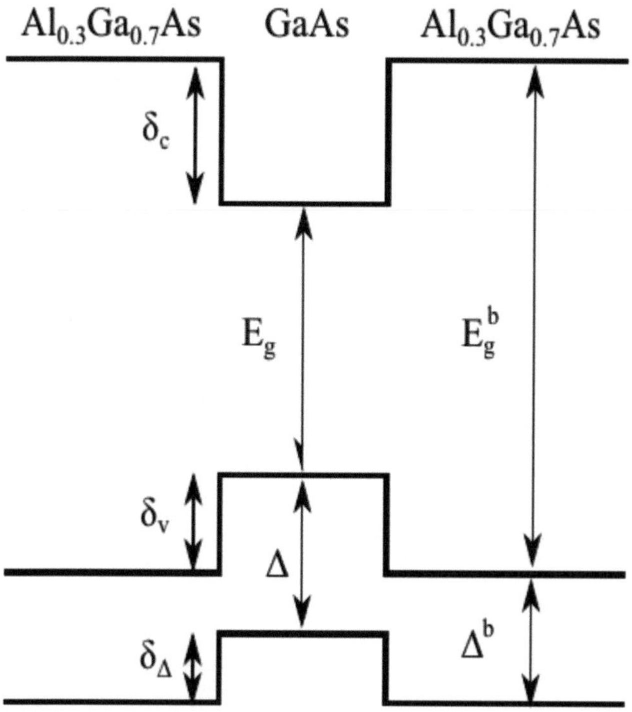

Figure 1.
A schematic of the band offsets for GaAs/AlGaAs well, in which E_g (E_g^b) and Δ (Δ^b) are the fundamental bandgap and the split-off gap in the well (barrier), respectively. δ_c, δ_v, and δ_Δ represent the band offsets between well and barrier layers, of the conduction band, heavy hole (and light hole), and split-off hole, respectively.

$$\mathcal{H}_{RD} = \frac{1}{2} g\mu_B \sum_{\nu=1,\,2} \left[\tau_\nu \otimes \boldsymbol{\sigma} \cdot \mathbf{B}_{SO}^{\nu} + \tau_{x_+} \otimes \boldsymbol{\sigma} \cdot \mathbf{B}_{SO}^{12}\right], \tag{5}$$

with g the electron g factor, μ_B the Bohr magneton, and $\tau_{1,2} = (\mathbf{1} \pm \tau_z)/2$. Explicitly, the intrasubband SO field is

$$\mathbf{B}_{SO}^{\nu} = \frac{2}{g\mu_B} k \left\{\left[\left(\alpha_\nu - \beta_{\nu,\,\mathrm{eff}}\right)\sin\theta - \beta_{\nu,\,3}\sin 3\theta\right]\hat{\mathbf{x}}_+ - \left[\left(\alpha_\nu + \beta_{\nu,\,\mathrm{eff}}\right)\cos\theta - \beta_{\nu,\,3}\cos 3\theta\right]\hat{\mathbf{x}}_-\right\}, \tag{6}$$

and the intersubband SO field is

$$\mathbf{B}_{SO}^{12} = \frac{2}{g\mu_B} k[(\eta - \Gamma)\cos\theta\hat{\mathbf{x}}_+ - (\eta + \Gamma)\sin\theta\hat{\mathbf{x}}_-], \tag{7}$$

where θ is the angle between $\mathbf{k}$ and the x_+ axis.

The Rashba and Dresselhaus SO coefficients, α_ν, β_ν, η, and Γ, appearing in Eqs. (6) and (7) read

$$\eta_{\nu\nu'} = \langle\psi_\nu|\eta_{\mathrm{w}}\partial_z V_{\mathrm{w}} + \eta_{\mathrm{H}}\partial_z\left(V_{\mathrm{g}} + V_{\mathrm{d}} + V_{\mathrm{e}}\right)|\psi_{\nu'}\rangle, \tag{8}$$

and

$$\Gamma_{\nu\nu'} = \gamma\langle\psi_\nu|k_z^2|\psi_{\nu'}\rangle, \tag{9}$$

with the Rashba coefficients $\alpha_\nu \equiv \eta_{\nu\nu}$, $\eta \equiv \eta_{12}$ and the Dresselhaus coefficients $\beta_\nu \equiv \Gamma_{\nu\nu}$ and $\Gamma \equiv \Gamma_{12}$. The coefficient $\beta_{\nu,\,\mathrm{eff}} = \beta_\nu - \beta_{\nu,\,3}$ (Eq. (6)) is the renormalized "linear" Dresselhaus coupling, due to the cubic correction $\beta_{\nu,\,3} = \gamma k_\nu^2/4$, where $k_\nu = \sqrt{2\pi n_\nu}$ is the νth-subband Fermi wave number with n_ν the νth-subband occupation.

Note that the Rashba strength α_ν (Eq. (8)) can be written in terms of several distinct "individual" contributions, i.e., $\alpha_\nu = \alpha_\nu^{\mathrm{g}} + \alpha_\nu^{\mathrm{d}} + \alpha_\nu^{\mathrm{e}} + \alpha_\nu^{\mathrm{w}}$, with $\alpha_\nu^{\mathrm{g}} = \eta_{\mathrm{H}}\langle\psi_\nu|\partial_z V_{\mathrm{g}}|\psi_\nu\rangle$ the gate contribution, $\alpha_\nu^{\mathrm{d}} = \eta_{\mathrm{H}}\langle\psi_\nu|\partial_z V_{\mathrm{d}}|\psi_\nu\rangle$ the doping contribution, $\alpha_\nu^{\mathrm{e}} = \eta_{\mathrm{H}}\langle\psi_\nu|\partial_z V_{\mathrm{e}}|\psi_\nu\rangle$ the electron Hartree contribution, and $\alpha_\nu^{\mathrm{w}} = \eta_{\mathrm{w}}\langle\psi_\nu|\partial_z V_{\mathrm{w}}|\psi_\nu\rangle$ the structural contribution. Similar for the intersubband Rashba term $\eta = \eta^{\mathrm{g}} + \eta^{\mathrm{d}} + \eta^{\mathrm{e}} + \eta^{\mathrm{w}}$, the matrix element for each contribution is calculated between different subbands. For convenience, below we use $\alpha_\nu^{\mathrm{g}+d} = \alpha_\nu^{\mathrm{g}} + \alpha_\nu^{\mathrm{d}}$ and $\eta^{\mathrm{g}+d} = \eta^{\mathrm{g}} + \eta^{\mathrm{d}}$ for the contribution from the gate plus doping potential. We should emphasize that all of the SO coupling contributions above depend on the total self-consistent potential V_{sc} as our wave functions are calculated self-consistently.

It is worth noting that here we do not consider in our model the many-body effect-induced discontinuity of the electron density upon occupation of the second subband, as demonstrated by Goni et al. [27] and Rigamonti and Proetto [28] at zero temperature. As this discontinuity vanishes for T>30 K, we believe this is a minor effect in our system when $T = 75$ K. However, it is conceivable that related features can manifest in the SO couplings at zero temperature. Additional work is needed to investigate this interesting possibility.

3. Two distinct regimes for the control of SO interaction

In this section, we first introduce the structure of our wells and relevant parameters adopted in our simulation. Then we discuss our calculated SO couplings for the two distinct regimes. In either regime, we focus on a well having a two-subband electron occupation at zero bias (i.e., $V_g = 0$). By tuning V_g, we alter electron occupations from two subbands to one subband. For each value of V_g, within the self-consistent Hartree approximation, we determine the relevant SO strengths, i.e., the intrasubband α_ν, $\nu = 1, 2$, and intersubband η Rashba couplings, and similarly for the Dresselhaus term, the intrasubband β_ν and the intersubband Γ. Finally, the case of the two regimes in between is discussed as well.

3.1 System

The quantum wells we consider are similar to the samples experimentally studied by Koralek et al. [7]: the 001-grown GaAs wells of width w sandwiched between 48-nm $Al_{0.3}Ga_{0.7}As$ barriers. Our structure contains only one delta-doping (Si) layer, sitting 17 nm away from the well interface, with donor concentration n_d. Our prior simulated SO couplings [26] on these samples agree well with the data obtained via the transient-spin-grating technique [7], where there is no bias applied and all wells share the same areal electron density n_e. Here we go beyond the experiment in [7] by using an external bias, which can vary the electron density and control the SO interaction. At zero bias we assume $n_e = n_d$ [7].

The width w of quantum wells we consider ranges from 20 to 70 nm. Two distinct regimes of the SO interaction we found are marked off around $w = w_c =$ 30–35 nm. Without loss of generality, we focus on a well of $w = 25$ nm for the first regime and a well of $w = 65$ nm for the second one. For the 25-nm well, we choose $n_d = 8.0 \times 10^{11}$ cm^{-2} and $T = 75$ K, the same as experimental parameters in [7], not only ensuring the electron occupation of the second subband at zero bias [26] but also retaining this second subband occupation over a broad range of V_g's. While for the 65-nm well, we choose relatively small values of n_d and T, i.e., $n_d = 4.0 \times 10^{11}$ cm^{-2} and $T = 0.3$ K, to exclude the electron occupation of a higher third subband.

3.2 Relevant parameters

In our GaAs/$Al_{0.3}Ga_{0.7}As$ wells, the values of relevant band parameters (see Eqs. (2) and (3)) are as follows: $E_g = 1.519$ eV and $\Delta = 0.341$ eV [4, 29, 30]. The corresponding counterparts of E_g and Δ in $Al_{0.3}Ga_{0.7}As$ barriers are $E_g^b = 1.951$ eV and $\Delta^b = 0.329$ eV [13, 24, 30]. As a consequence, we have the band offsets for conduction and valence bands, $\delta_c = \left(E_g^b - E_G\right) \times 60.4\% = 0.261$ eV, $\delta_v = E_g^b - E_G - \delta_c = 0.171$ eV, and $\delta_\Delta = \delta_v - \Delta + \Delta^b = 0.159$ eV [13, 24, 30]. For the Kane parameter, we choose $P = 1.0493$ nm [4, 29]. Then, we obtain $\eta_w = 3.97$˚A2 and $\eta_H = -5.30$˚A2 (Eqs. (2) and (3)). We treat the bulk Dresselhaus constant γ as an empirical parameter. We have recently done detailed calculations on a set of GaAs wells and have found via a realistic fitting procedure (theory and experiment) $\gamma \sim 11.0$ eV˙˚A3 [13]. We use this value in our simulations, consistent with the one obtained in a recent study by Walser et al. [31].

3.3 Numerical outcome: two distinct regimes

Below we discuss our self-consistent outcome for the SO couplings. We present our calculated intra- and intersubband SO couplings in the two distinct regimes. The behavior of the SO interaction in the first regime as a function of the gate voltage is usual. As a consequence, we mainly focus on the second regime, in which new features of the SO interaction emerge.

3.3.1 Intrasubband SO couplings: both Rashba and Dresselhaus terms

We consider in the first regime a well of $w = 25$ nm. In **Figure 2(b)**, we show the dependence of Rashba α_ν ($\nu = 1, 2$) and Dresselhaus β_ν coefficients of the two subbands on V_g. We find that α_1 and α_2 have the same sign and they both decrease to zero near the symmetric configuration (at $V_g \sim 0.36$ eV) (see arrow in **Figure 2(b)**) and further change their sign. For the confining electron potential V_{sc} and wave functions ψ_ν in this symmetric geometry of the well, see **Figure 2(a)**. Clearly, our wells are asymmetric at $V_g = 0$ because of the one-side delta doping. In contrast to the Rashba term, we find that the Dresselhaus couplings $\beta_\nu = \gamma\langle\psi_\nu|k_z^2|\psi_\nu\rangle$ remain essentially constant with V_g. The inset in **Figure 2(b)** shows the electron density n_ν of the two subbands with $n_1 + n_2 = n_e$ as a function of V_g, and the second subband starts to be unpopulated around $V_g \sim 0.5$ eV.

Now, we turn to the second regime, in which we consider a well of $w = 65$ nm, as shown in **Figure 2(c)**. As opposed to the first regime, we find that α_1 and α_2 could have opposite signs. This is because electrons occupying the first and second subbands tend to be distributed on opposite sides for a wide well, as a result of the

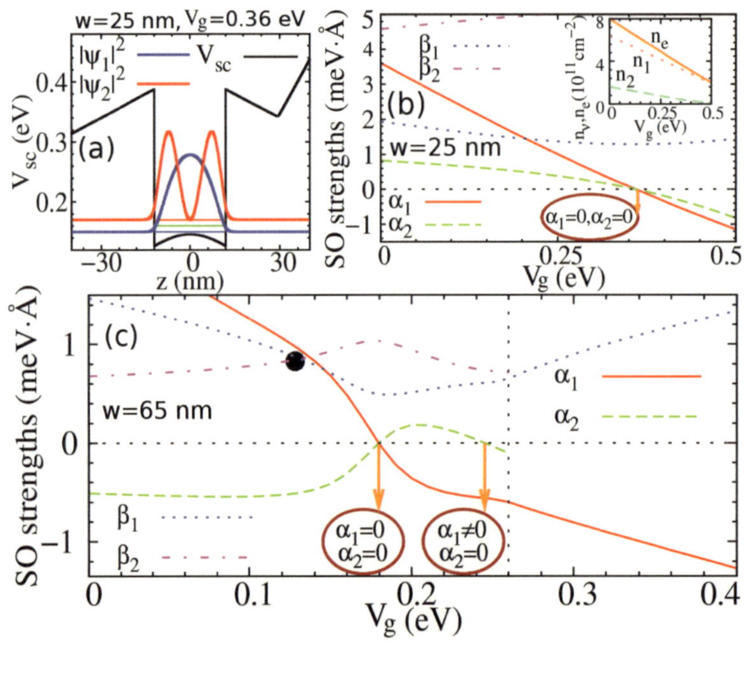

Figure 2.
(a) Self-consistent potential V_{sc} and wave function profiles ψ_ν of a 25-nm well at $V_g = 0.36$ eV (symmetric configuration). The horizontal lines inside the well respectively indicate the two-subband energy levels and the Fermi level. Gate control of Rashba α_ν and Dresselhaus β_ν SO couplings, for the 25-nm (b) and 65-nm wells (c). In (b), the inset shows electron density n_ν with $n_1 + n_2 = n_e$ as a function of V_g. In (c), the vertical dashed line at $V_g \sim 0.26$ eV marks regions of one- and two-band occupations. The temperature is chosen at 75 K for the 25-nm well and 0.3 K for the 65-nm well, from Wang et al. [32].

electron Hartree potential created "central barrier"; see **Figure 3(a)–(c)**. We emphasize that the sign of SO couplings is crucial in diverse subjects of the field of spintronics, e.g., the persistent skyrmion lattice [25], the nonballistic spin field transistor operating with orthogonal spin quantization axes [33], and the transition from a topological insulator to Dirac semimetal [34].

To see more details about the sign change of Rashba couplings, we show in **Figure 3(d)** and **(e)** the gate dependence of distinct contributions of α_1 and α_2 separately, i.e., α_ν^e, α_ν^w, and α_ν^{g+d}. For the structural contribution,

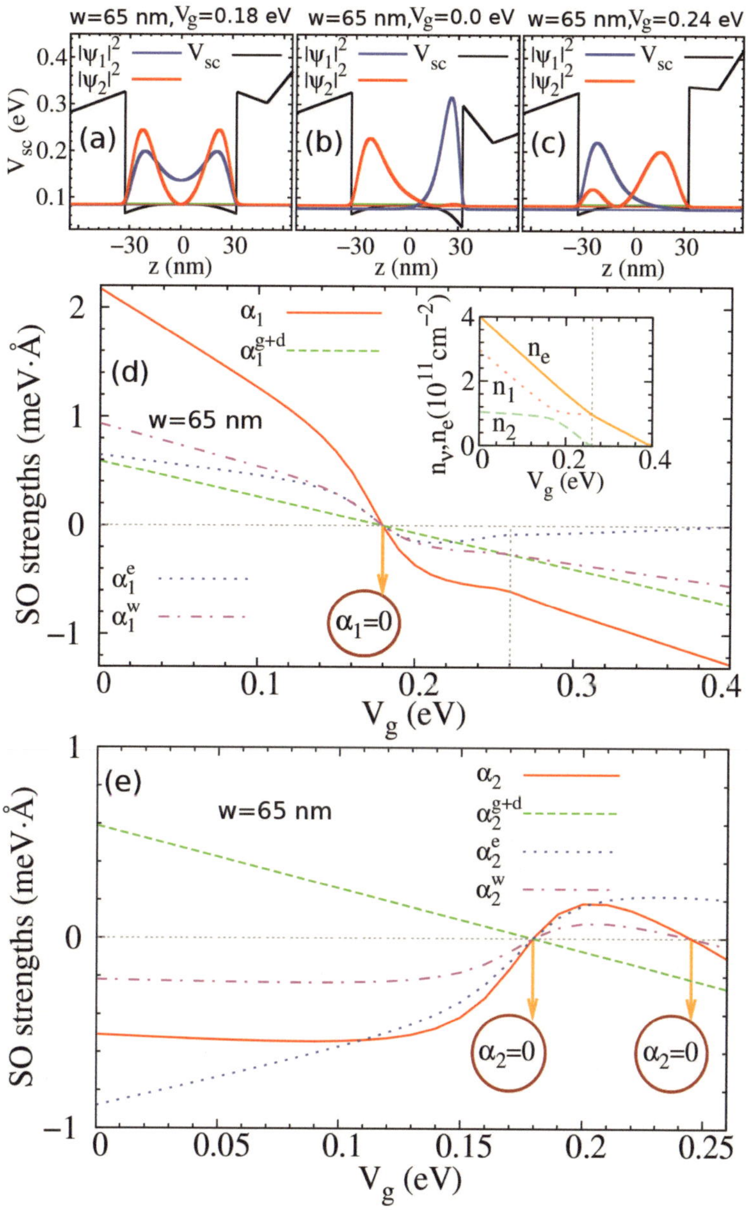

Figure 3.
(a)–(c) Potential V_{sc} and wave function profiles ψ_ν ($\nu = 1, 2$) for a 65-nm well at three values of voltages, with $V_g = 0.18$ (a), 0 (b), and 0.24 eV (c). The horizontal lines inside the well indicate the two-subband energy levels and the Fermi level. Dependence of different contributions to the intrasubband Rashba strength of the first (d) and second (e) subbands on the gate voltage for the 65-nm well. In (d), the inset shows the electron density n_ν with $n_1 + n_2 = n_e$ as a function of gate voltage. The vertical dashed line in (d) (and inset) marks regions of one- and two-band occupations. The temperature is chosen at 0.3 K, from Wang et al. [32].

$\alpha_\nu^{w} \propto \langle \psi_\nu | \partial_z V_w | \psi_\nu \rangle = |\psi_\nu(w/2)|^2 - |\psi_\nu(-w/2)|^2$ [26], α_1^{w} is dominated by the first term, while α_2^{w} is dominated by the second term, due to the already discussed tendency of the subband wave functions of ψ_1 and ψ_2 to respectively localize on the right and left halves of the well (zero bias for instance). This leads to opposite signs between α_1^{w} and α_2^{w}. Similarly, for $\alpha_\nu^{e} \propto \langle \psi_\nu | \partial_z V_e | \psi_\nu \rangle$, the electron Hartree contributions of the two subbands have opposite signs as well since V_e mostly has opposite slopes with respect to z for $z>0$ and $z<0$ [24]. However, for $\alpha_\nu^{g+d} \propto \langle \psi_\nu | \partial_z (V_g + V_d) | \psi_\nu \rangle$, the corresponding contributions of the two subbands have the same sign, due to the fact that the gate plus doping potential is linear across the well region. The relative sign of the total contribution to α_1 and α_2 depends on the interplay of the three constituents as a function of gate voltages.

From **Figure 3(e)**, we also find that α_2 remains essentially constant with V_g at lower voltages, precisely, as its constituent α_2^{w} behaves except for a prefactor difference. In contrast, it is found that α_1 sensitively depends on V_g in the whole parameter range studied here. Moreover, α_1 and α_2 become zero (see arrow in **Figure 3(d)** and left arrow in **Figure 3(e)**) across the symmetric configuration, corresponding to $V_g \sim 0.18$ eV. Interestingly, we observe that α_2 can even attach zero at $V_g \sim 0.24$ eV (see right arrow in **Figure 3(e)**), at which the well is very asymmetric (**Figure 3(c)**).

Before moving into the Dresselhaus couplings in this second regime, it is worth noting that the electron densities of the two subbands exhibit the anticrossing-like behavior near the symmetric configuration (at $V_g \sim 0.18$ eV), as shown in the inset of **Figure 3(d)**. The feature of anticrossing of electron densities has been experimentally observed in double wells by Fletcher et al. [35], thus implying that our wells in the second regime do behave like *effective* double wells.

In **Figure 2(c)**, we show the linear Dresselhaus couplings $\beta_\nu = \gamma \langle \psi_\nu | k_z^2 | \psi_\nu \rangle$ in the second regime. We observe that the inequality $\beta_1 < \beta_2$, which is valid in the first regime, only holds in configurations near the symmetric geometry. However, if the well is very asymmetric, i.e., at $V_g = 0$, we find $\beta_1 > \beta_2$ [see crossing (black dot) between β_1 and β_2 in **Figure 2(c)**], in contrast to the first regime.

Now, we are ready to determine the persistent-spin-helix (PSH) symmetry points of the two subbands, at which the Rashba α_ν and the renormalized linear Dresselhaus $\beta_{\nu,\mathit{eff}}$ (due to cubic corrections) are equal in strength, i.e., $\alpha_\nu \approx \beta_{\nu,\mathit{eff}}$, as shown in **Figure 4**. We observe that $\alpha_1 \approx \beta_{1,\mathit{eff}}$ takes place at $V_g \sim 0.17$ eV (see right arrow in **Figure 4**) for the first subband, while $\alpha_2 \approx -\beta_{2,\mathit{eff}}$ always holds over the range of $V_g = 0 - 100$ meV (see left arrow in **Figure 4**) for the second subband. This is possibly facilitating the locking of the PSH symmetry in practice. Note that in contrast to the well matched $|\alpha_2|$ and $\beta_{2,\mathit{eff}}$ here, a deviation between them can possibly occur for other wells with strong SO couplings (e.g., InAs-based wells). However, this deviation can always be possibly overcome by properly varying the electron density, which controls the cubic $\beta_{2,3}$—thus tuning $\beta_{2,\mathit{eff}}$ to match with $|\alpha_2|$. These symmetry points are also crucial for the persistent skyrmion lattice [25] as well as the nonballistic spin field effect transistors [5] operating with orthogonal spin quantization axes [33].

We emphasize that, for the PSH symmetry points that we determined above, the effect of the interband SO couplings (see Section 3.3.2) and of the random Rashba coupling [36–39] has been ignored. For the former, it is only relevant near the crossing(s) of the two-subband branches, as discussed in [25]. For the latter, it may in general destroy the helix but has a negligible effect on the results for our wells here [26].

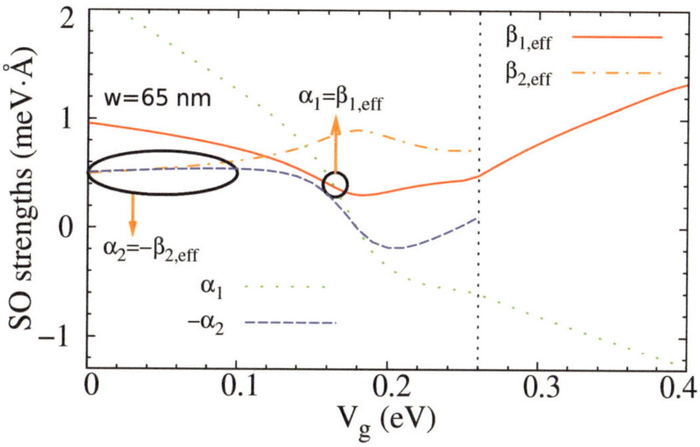

Figure 4.
The dependence of intrasubband Rashba α_ν and renormalized Dresselhaus $\beta_{\nu,\,eff}$ coefficients on V_g for the 65-nm well. The vertical dashed line marks regions of one- and two-band occupations. The temperature is held fixed at 0.3 K, from Wang et al. [32].

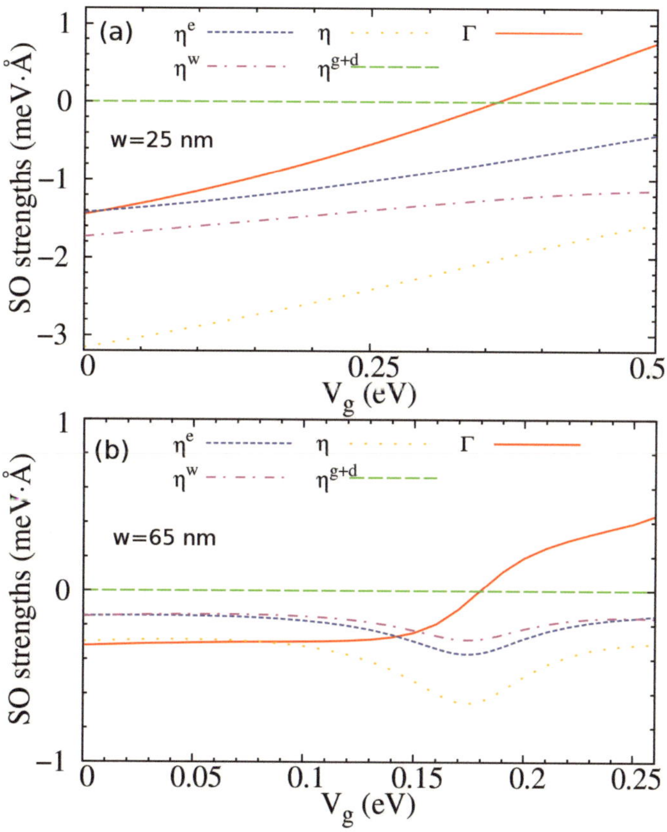

Figure 5.
Gate control of intersubband Rashba η and Dresselhaus Γ strengths for a 25-nm well (a) and a 65-nm well (b) the temperature is chosen at 75 K for the 25-nm well and 0.3 K for the 65-nm well, from Wang et al. [32].

3.3.2 Intersubband SO couplings: both Rashba and Dresselhaus terms

Below we turn to the interband SO terms. Referring to the first regime, in **Figure 5(a)**, we show the intersubband Rashba coupling η and its distinct constituents including η^{g+d}, η^{e}, and η^{w} and Dresselhaus coupling Γ for a well of $w = 25$ nm. We find that both η and Γ exhibit basically the linear behavior with V_g. Owing to the orthogonality between ψ_1 and ψ_2, the gate plus doing contribution η^{g+d} vanishes. As for the constituents η^{w} and η^{e}, the dominant variation with the gate is the latter as the electron density in wells changes with V_g, even though the magnitude of η^{w} is greater than η^{e}.

For the second regime, we show in **Figure 5(b)** the intersubband SO couplings for a well of $w = 65$ nm. In contrast to the first regime, here η largely remains constant as V_g increases. And a maximum of $|\eta|$ occurs at $V_g \sim 0.18$ eV (symmetric configuration). This arises from the fact that the wave functions of the two subbands, ψ_1 and ψ_2, are mostly separated in very asymmetric configurations (**Figure 3(b)** and **(c)**), which renders η (depending on the overlap of ψ_1 and ψ_2) almost constant and relatively weak. Similar to η, Γ also weakly depends on the gate except for configurations of the well near the symmetric geometry. Finally, we emphasize that, for wide and asymmetric *enough* wells, where ψ_1 and ψ_2 have vanishing overlap, both η and Γ tend to be zero.

3.4 Two regimes in between for the control of SO couplings

Now, it is clear that the SO couplings show distinct behaviors for the two regimes. By analyzing SO couplings for a set of wells of $w = 20 - 70$ nm, below we

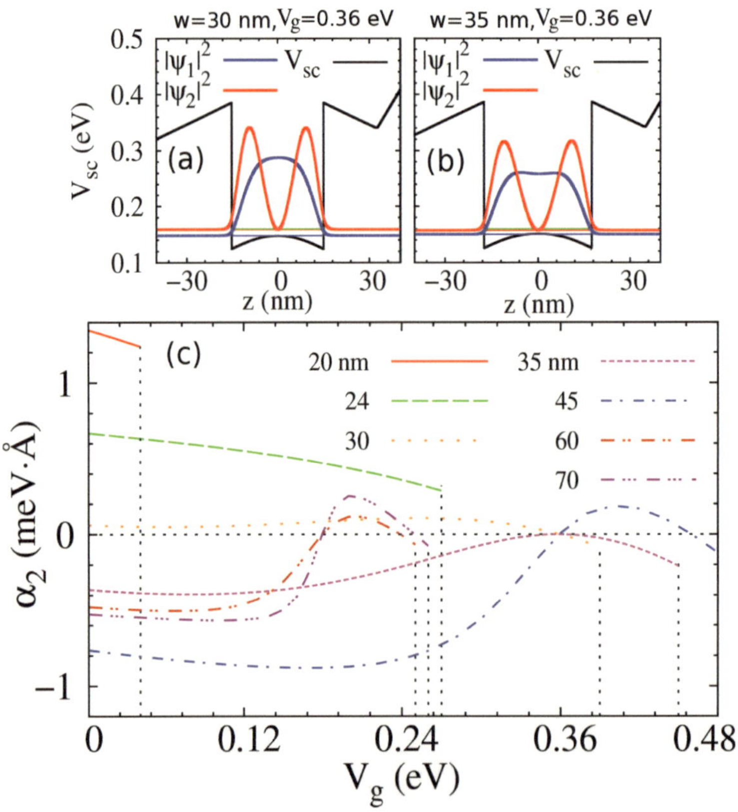

Figure 6.
Confining potential V_{sc} and wave function profiles ψ_ν in GaAs wells at $V_g \sim 0.36$ eV (symmetric geometry), for a 30-nm well (a) and a 35-nm well (b). the horizontal lines inside the wells indicate the two-subband energy levels and the Fermi level (c). the dependence of intrasubband Rashba α_2 of the second subband on V_g, for a set of wells of w = 20,24,30,35,45,60, and 70 nm. The vertical dashed lines mark regions of one- and two-band occupations. The temperature is held fixed at 0.3 K, from Wang et al. [32].

discuss how the two regimes transit from one to the other. To explore this, we only focus on α_2, which is mostly in contrast between the two regimes.

In **Figure 6**(**c**), we show the dependence of α_2 on V_g. Clearly, for the 20- and 24-nm wells, α_2 changes almost linearly, showing the usual behavior of the Rashba coupling, as demonstrated in the first regime. However, for the wells of $w = 45, 60$, and 70 nm, we observe the unusual behavior, which is referred to the second regime. When w in the range of $w = 30 - 35$ nm, the gate dependence of α_2 behaves in between the two regimes, i.e., there is neither the linear behavior in the first regime nor the sign of double changes in the second one. To gain more insight into this, we examine the confining potential V_{sc} and wave function profiles ψ_ν for the wells of $w = 30$ and 35 nm (**Figure 6**(**a**) and (**b**)), which are used to determine the SO coupling. It is clear that the envelope wave functions for the 30- and 35-nm wells look like a bridge connecting the two regimes, cf. **Figure 2**(**a**) for the first regime, **Figure 3**(**a**) for the second regime, and **Figure 6**(**a**) and (**b**) for the two regimes in between.

To deplete the second subband occupation, it is clear that a wider well requires in general a larger value of gate voltage (see vertical dashed lines for wells of $w = 20 - 45$ nm), as the confinement becomes weaker for a wider well. However, for the 60- and 70-nm wells (see vertical dashed lines for wells of $w = 60 - 70$ nm), we find that the voltages needed to deplete the second subband occupation are even less than the one for the 24-nm well. This is attributed to the low electron density we choose for wide wells, the choice of which is to exclude the electron occupation of a higher third subband.

4. Control of SO interaction from single to double and triple wells

With the knowledge of the SO interaction in single wells (Section 3), below we consider the case of multiple wells and determine the electrical control of the SO interaction in GaAs from single to double and triple wells.

4.1 System

The main structure of our well is again similar to the samples experimentally studied by Koralek et al. [7]: the 001-grown GaAs well of width $w = 26$ nm sandwiched between 48-nm $Al_{0.3}Ga_{0.7}As$ barriers. To have double (triple) wells, we introduce one (two) additional $Al_xGa_{1-x}As$ barrier layers of length $b = 3$ nm embedded inside the above structure. We choose the total electron density $n_e = 6.0 \times 10^{11}$ cm^{-2}. The temperature is held fixed at $T = 0.3$ K. Note that we choose relatively low n_e and T, instead of the experimental values $n_e = 8.0 \times 10^{11}$ cm^{-2} and $T = 75$ K in [7], on the one hand, to ensure the second subband electron occupation and on the other hand, to exclude the electron occupation of a higher third subband, for all values of V_gs studied here.

4.2 SO coupling coefficients

To explore the SO features from single to double and triple wells, firstly, we focus on the case of having only one $Al_xGa_{1-x}As$ additional barrier embedded at the center of the system.

In **Figure 7**(**a**) and (**b**), we show the gate dependence of intrasubband Rashba terms in our $GaAs/Al_{0.3}Ga_{0.7}As$ wells, for several Al contents x of the central barrier $Al_xGa_{1-x}As$. At $x = 0$ (single well), we find the usual scenario, i.e., α_1 and α_2 have

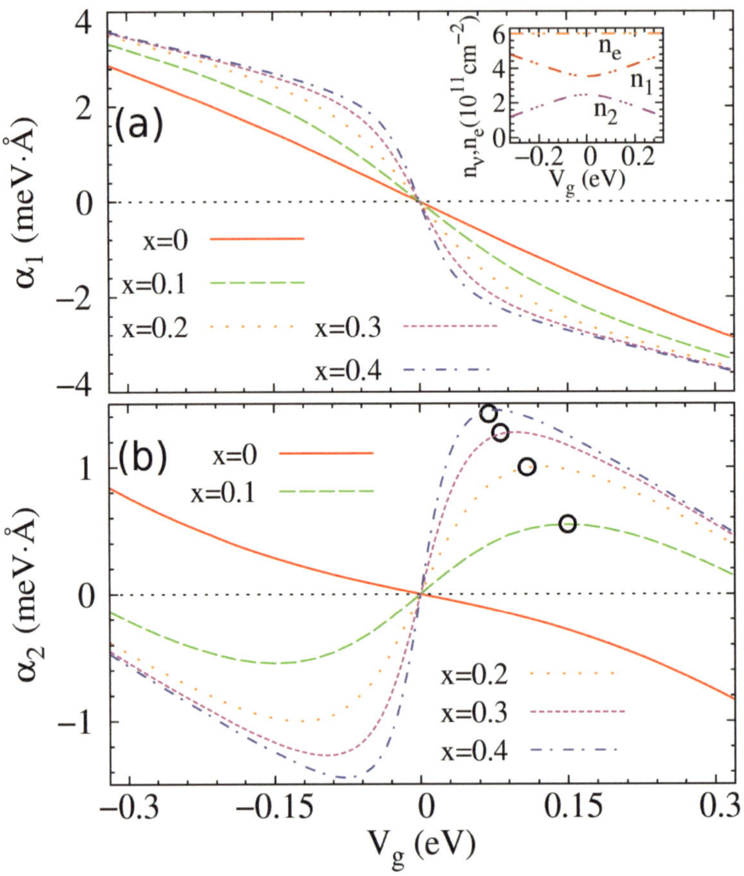

Figure 7.
Intrasubband Rashba α_1 (a) and α_2 (b) couplings as a function of V_g in GaAs/$Al_{0.3}Ga_{0.7}As$ wells for several Al contents x of a central barrier $Al_xGa_{1-x}As$. In (a), the inset shows the gate dependence of electron occupations n_1, n_2, and $n_e = n_1 + n_2$. In (b), the black circles indicate where α_2 exhibits the maximum, from Wang et al. [40].

the same sign and both change almost linearly with V_g. In this case, electrons occupying the first and second subbands spread over the whole well region (**Figure 8(a)**) and "see" the same fields (i.e., slope of potentials) that determine Rashba couplings (Eq. (8)). Thus, the above behavior of α_ν with V_g follows. In contrast, when x is away from zero (double well), electrons of the two subbands tend to localize on the left and right sides of the well, respectively, due to the effect of the central barrier (**Figure 8(b)** and **(c)**). This results in the opposite signs of α_1 and α_2, as the local field on the left and right sides of the well is almost reversed; see potential profiles in wells shown in **Figure 8(b)** and **(c)**. In addition, α_2 first increases with V_g, while peaks at some point depending on x and further decreases, following from the compensated interplay of *local* fields (i.e., electron Hartree plus structural well) and the *universal* external gate field. For a larger value of x, α_2 increases with V_g more abruptly before it peaks, see black circles in **Figure 7(b)** for a set of values of x. Note that α_1 in magnitude consistently increases with V_g, since the local fields and the universal external gate field contribute to α_1 constructively.

Figure 8(c) shows the dependence of linear intrasubband Dresselhaus coupling $\beta_\nu = \gamma\langle k_z^2\rangle$ on V_g. As opposed to α_ν, we find that β_ν is weakly dependent on V_g ($\pm \sim 0.5$ meV Å) for all values of x considered here. On the other hand, as x increases, the coupling between the left and right sides of the well is getting weaker,

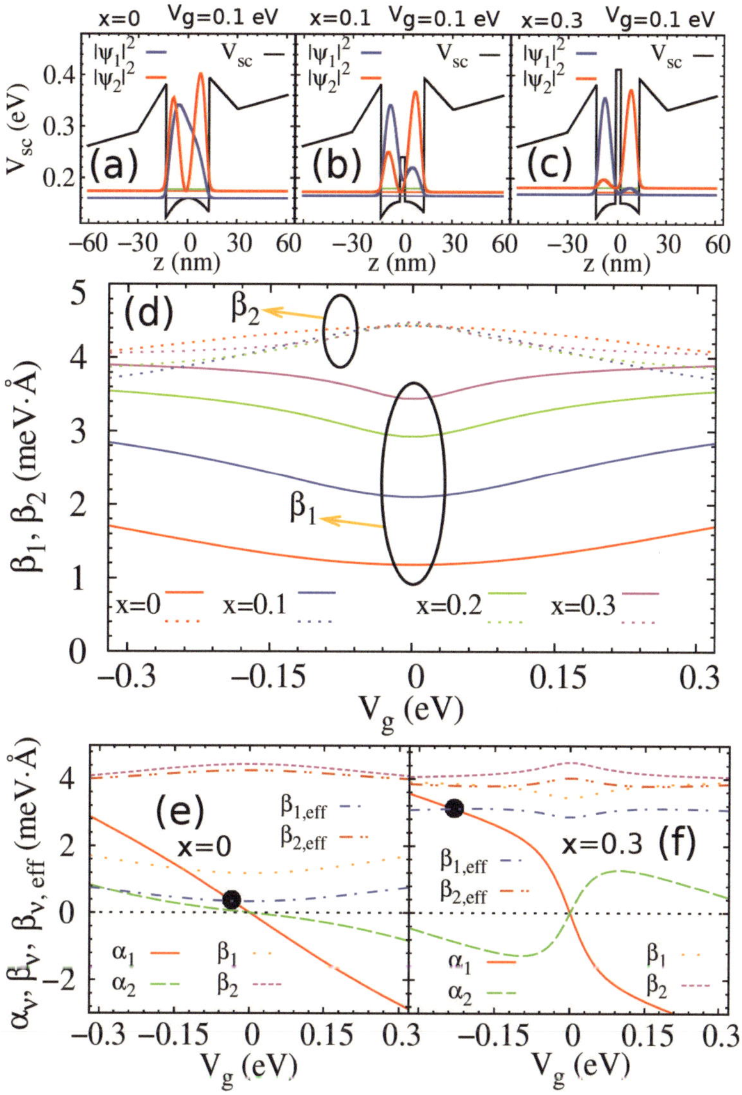

Figure 8.
(a)–(c) Self-consistent potential V_{sc} and wave function profiles ψ_ν for a GaAs/$Al_{0.3}Ga_{0.7}As$ well with $Al_xGa_{1-x}As$ as a central barrier, with $x = 0$ (a), 0.1 (b), and 0.3 (c). We consider $V_g = 0.1\,eV$. The horizontal lines inside the well indicate the two-subband energy levels and the Fermi level. (d) Intrasubband Dresselhaus terms β_ν as a function of V_g, for several values of xs with $x = 0, 0.1, 0.2, 0.3$. (e) and (f) Gate dependence of SO strengths (α_ν, β_ν, $\beta_{\nu,\text{eff}}$) for $x = 0$ and $x = 0.3$, respectively. The black points indicate the PSH symmetry points (i.e., $\alpha_\nu = \pm\beta_{\nu,\text{eff}}$), from Wang et al. [40].

leading to a shrinking of the energy separation of the two subbands. Consequently, the strength of β_1 and β_2 becomes more close for a larger value of x.

Now, we determine the PSH symmetry points, where the Rashba α_ν and the renormalized linear Dresselhaus $\beta_{\nu,\text{eff}} = \beta_\nu - \beta_{\nu,3}$ are matched. The cubic Dresselhaus term $\beta_{\nu,3} = \gamma k_\nu^2/4 = \gamma\pi n_\nu/2$ depends on the electron occupations and therefore alters the condition of the PSH symmetry. **Figure 8(e)** and **(f)** show the gate dependence of α_ν, β_ν, and $\beta_{\nu,\text{eff}}$, for $x = 0$ and 0.3, respectively. Clearly, we see the distinction between β_ν and $\beta_{\nu,\text{eff}}$ because of the correction from $\beta_{\nu,3}$, in particular for the first subband, which has a higher electron occupation (see inset of **Figure 7(a)**). In the parameter range considered here, we only attain the PSH symmetry for the first subband ($\alpha_1 = \beta_{1,\text{eff}}$); see black point in **Figure 8(e)** and **(f)**.

To attain the PSH symmetry for the second subband, one needs an even wider well [26], where β_2 is weaker so that α_2 and β_2 have the feasibility of matched strength by tuning V_g.

Besides altering the PSH symmetry condition involving SO terms of the first harmonic (sin / cos functions of θ), the cubic Dresselhaus $\beta_{3,\nu}$ also has the third harmonic (3θ) contribution; see Eq. (6). This leads to the decay of the PSH due to the D'yakonov-Perel' (DP) spin relaxation mechanism [41]. Specifically, the third harmonic Dresselhaus coupling-induced spin relaxation rate at the PSH symmetry point is written as $\Gamma_D^\nu = \gamma^2 k_\nu^6 \tau_P / 4\hbar^2$ [42], with τ_P the momentum relaxation time. For GaAs wells, with $\gamma \sim 11.0$ eV°A3, $k_\nu \sim 0.2\ \text{nm}^{-1}$, and $\tau_P \sim 1.0$ ps [13, 31], we find $\Gamma_D^\nu \sim 0.07\ \text{ns}^{-1}$.

In **Figure 9(a)** and **(b)**, we show the gate dependence of the intersubband Rashba η and Dresselhaus Γ couplings, which depend on the overlap of the wave functions of the two subbands; see Eqs. (8) and (9). For the intersubband Rashba coupling η, we find that it remains essentially constant at $x = 0$, since ψ_1 and ψ_2 are well overlapped even at asymmetric configurations; see **Figure 8(a)** with $V_g = 0.1$ eV. However, when x is nonzero, electrons of the two subbands have the tendency of localizing on opposite sides of the well at asymmetric configurations (**Figure 8(b)** and **(c)**), namely, ψ_1 and ψ_2 tend to be separated. Therefore, η exhibits the strongest at the symmetric configuration of the system (i.e., $V_g = 0$). Note that at the symmetric configuration, ψ_1 and ψ_2 are perfectly overlapped. Furthermore, the separation of ψ_1 and ψ_2 is more distinct for a larger x (same V_g), cf. **Figure 8(b)** and **(c)**,

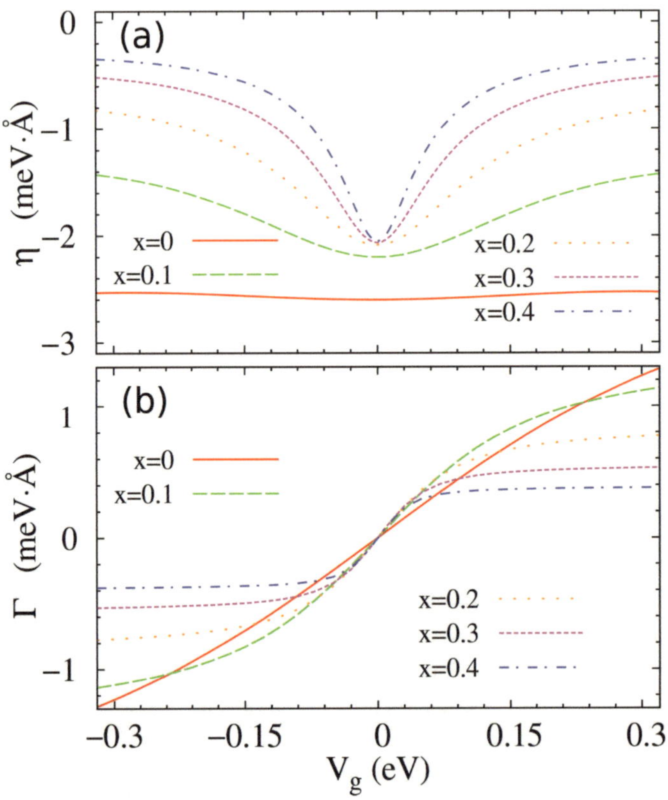

Figure 9.
Intersubband Rashba η (a) and Dresselhaus Γ (b) couplings as a function of V_g in GaAs/$Al_{0.3}Ga_{0.7}As$ wells, for several Al contents x of the central barrier $Al_xGa_{1-x}As$ with $x = 0, 0.1, 0.2, 0.3, 0.4$, from Wang et al. [40].

which renders η more sensitively dependent on V_g near $V_g = 0$. As for the intersubband Dresselhaus term $\Gamma = \langle 1|k_z^2|2\rangle$, on the one hand, it depends on the overlap of ψ_1 and ψ_2 and accordingly strongly depends on V_g near $V_g = 0$, similar to η. On the other hand, as opposed to η, the term Γ vanishes at $V_g = 0$ and further changes sign across the symmetric configuration, similar to the intrasubband Rashba terms α_ν.

Finally, we consider the case of our system having two additional barriers, namely, a triple well. As compared to the double well case, the wave functions of the two subbands ψ_1 and ψ_2 are more overlapped in a triple well under the same external conditions (e.g., same V_g), cf. **Figure 10(a)** and **(b)**, due to the electron distribution in the middle of three wells. As a consequence, the relatively smooth change of SO terms even near the symmetric configuration follows when V_g varies, as shown in **Figure 10(c)**. Moreover, we find that the Dresselhaus term β_ν is stronger than that in our double well, cf. **Figures 8(d)** and **10(c)**. Note that the basic feature of the electrical control of the SO interaction in our triple well is similar to that in the double well, for the parameter range studied, under which the two-subband electron occupancy occurs.

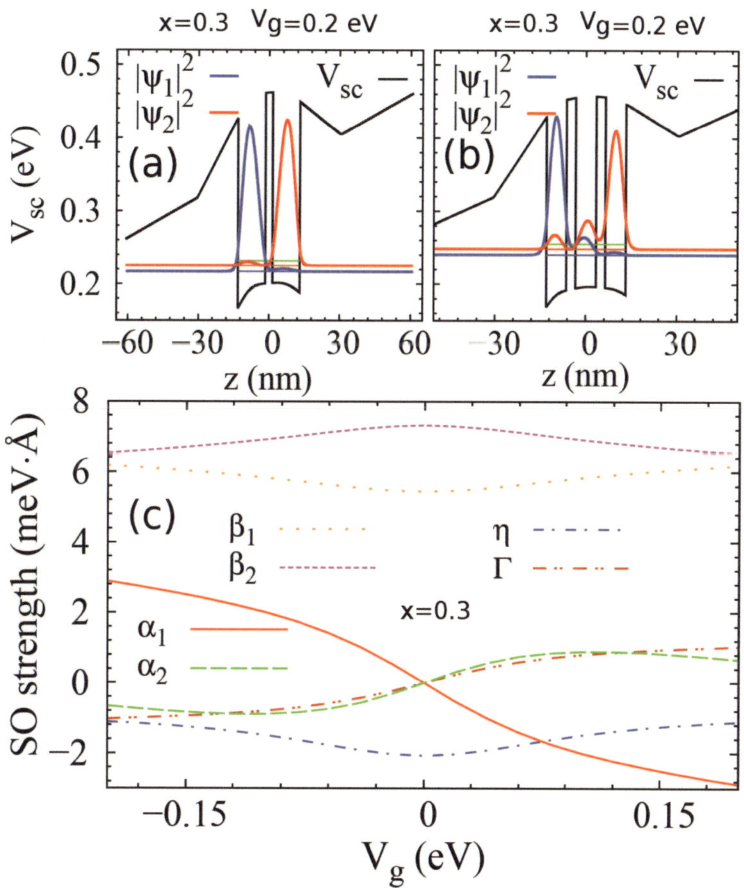

Figure 10.
Confining potential and the two-band wave function for a GaAs/$Al_{0.3}Ga_{0.7}As$ double well (a) and triple well (b) with embedded barrier $Al_xGa_{1-x}As$ of $x = 0.3$, at $V_g = 0.2$ eV. The horizontal lines inside the well indicate the two-subband energy levels and the Fermi level (c). gate control of intrasubband Rashba (α_ν) and Dresselhaus (β_ν) and Intersubband Rashba (η) and Dresselhaus (Γ) for the GaAs/$Al_{0.3}Ga_{0.7}As$ triple well with the embedded barrier $Al_xGa_{1-x}As$ of $x = 0.3$, from Wang et al. [40].

5. Conclusion

In this chapter, firstly, we consider two distinct regimes of the control of the SO interaction in conventional semiconductor quantum wells. Specifically, we have performed a detailed self-consistent calculation on realistic GaAs wells with gate-altered electron occupations from two subbands to one subband, thus determining how the SO coupling (both the magnitude and sign) changes as a function of the gate voltage V_g. We have considered a set of wells of the width w in a wide range and found two distinct regimes. In the first regime, the behavior of the SO interaction is usual, e.g., a linear gate control of the Rashba terms. In contrast, in the second regime, there are emerging new features that one has to pay attention to when controlling SO couplings, e.g., the relative signs (same or opposite) of α_1 and α_2 can be controlled by the gate, α_2 can attain zero in certain asymmetric configurations, and α_2 remains essentially constant within a particular gate voltage range. In addition, we have determined the persistent-spin-helix symmetry points of the two subbands and found that the condition $\alpha_2 = -\beta_{2,eff}$ always holds over a broad range of V_gs, thus possibly facilitating the locking of the symmetry point for the second subband in practice. Some of these features in the second regime can in principle be observed in proper double-well structures [24], as wide wells behave like "effective" double wells owing to the electron Hartree potential-induced central barrier. Moreover, the "symmetric configuration" mentioned throughout this work only refers to regions near the well (i.e., not far into the barriers). We cannot make our wells universally/fully symmetric by only tuning the gate voltage, due to the one-side doping in our system. This *partial* symmetric configuration is enough to render the intrasubband Rashba couplings to zero, since the envelope wave functions decay very quickly into the barriers. Our results should be timely and important for experiments controlling/tailoring the SO coupling *universally*, particularly for the *unusual* electrical control of the SO coupling in the second regime.

Secondly, we have investigated the full scenario of the electrical control of the SO interaction in a realistic GaAs/$Al_{0.3}Ga_{0.7}As$ well with one or two additional $Al_xGa_{1-x}As$ barriers embedded, in the course of the transition of our system from single to double and triple wells. We constantly consider the two-subband electron occupancy for all values of gate voltage V_g studied here. As the Al content of the embedded barrier(s) x varies, we find distinct scenarios of the electrical control of SO terms, e.g., linear or nonmonotonic dependence of α_2 on V_g, same or opposite signs between α_1 and α_2, and inert or abrupt change of η with V_g near the symmetric configuration. In addition, we find that the gate dependence of SO terms is more smooth and β_ν is more stronger in our triple well, compared to the double well case. Moreover, we observe that the basic scenario of the electrical control of the SO interaction in our triple and double wells is similar, in the parameter range studied here. These results are expected to be important for a broad control of the SO interaction in semiconductor nanostructures.

As a final remark, in the case of three-subband electron occupancy which is not considered here, the electrical control of SO couplings is possibly distinct between double and triple wells because of a higher third subband occupation. More work is needed to investigate this interesting possibility (higher electron density).

Acknowledgements

This work was supported by the National Natural Science Foundation of China (Grant No. 11874236), FAPESP, and Capes.

Author details

Jiyong Fu[1]*, Wei Wang[2] and Minghua Zhang[2]

1 Department of Physics, Qufu Normal University, Qufu, Shandong, China

2 Department of Physics, Jining University, Qufu, Shandong, China

*Address all correspondence to: jiyongfu78@gmail.com

References

[1] Awschalom D et al. Semiconductor Spintronics and Quantum Computation. New York: Springer; 2002

[2] Žutić I et al. Spintronics: Fundamentals and applications. Reviews of Modern Physics. 2004;**76**:32

[3] Datta S, Das B. Electronic analog of the electro-optic modulator. Applied Physics Letters. 1990;**56**:665

[4] Winkler R. Spin-Orbit Coupling Effects in Two-Dimensional Electron and Hole Systems. Berlin and New York: Springer; 2003

[5] Schliemann J et al. Nonballistic spin-field-effect transistor. Physical Review Letters. 2003;**90**:146801

[6] Bernevig BA et al. Exact SU(2) symmetry and persistent spin helix in a spin-orbit coupled system. Physical Review Letters. 2006;**97**:236601

[7] Koralek JD et al. Emergence of the persistent spin helix in semiconductor quantum wells. Nature. 2009;**458**:610

[8] Walser MP et al. Direct mapping of the formation of a persistent spin helix. Nature Physics. 2012;**8**:757

[9] Bernevig BA et al. Quantum spin hall effect and topological phase transition. Science. 2006;**314**:1757

[10] Lutchyn RM et al. Majorana fermions and a topological phase transition in semiconductor-superconductor heterostructures. Physical Review Letters. 2010;**105**: 077001

[11] Oreg Y et al. Helical liquids and majorana bound states in quantum wires. Physical Review Letters. 2010; **105**:177002

[12] Dresselhaus G. Spin-orbit coupling effects in zinc blende structure. Physics Review. 1955;**100**:580

[13] Dettwiler F et al. Stretchable persistent spin helices in GaAs quantum wells. Physical Review X. 2017;7:031010

[14] Bychkov YA, Rashba EI. Properties of a 2D electron gas with lifted spectral degeneracy. JETP Letters. 1984;**39**:78

[15] Engels G et al. Exact exchange plane-wave-pseudopotential calculations for slabs: Extending the width of the vacuum. Physical Review B. 1997;**55**:R1958

[16] Nitta J et al. Gate control of spin-orbit interaction in an inverted In0.53Ga0.47As/In0.52Al0.48As heterostructure. Physical Review Letters. 1997;**78**:1335

[17] Sasaki A et al. Direct determination of spin–orbit interaction coefficients and realization of the persistent spin helix symmetry. Nature Nanotechnology. 2014;**90**:703

[18] Hernandez FGG et al. Observation of the intrinsic spin hall effect in a two-dimensional electron gas. Physical Review B. 2013;**88**:161305(R)

[19] Bentmann H et al. Direct observation of interband spin-orbit coupling in a two-dimensional electron system. Physical Review Letters. 2012;**108**:196801

[20] Hu CM et al. Zero-field spin splitting in an inverted In0.53Ga0.47As/ In0.52Al0.48As heterostructure: Band nonparabolicity influence and the subband dependence. Physical Review B. 1999;**60**:7736

[21] Kavokin KV et al. Spin-orbit terms in multi-subband electron systems: A bridge between bulk and

two-dimensional Hamiltonians. Semiconductors. 2008;**42**:989

[22] de Andrada e Silva EA et al. Spin-orbit splitting of electronic states in semiconductor asymmetric quantum wells. Physical Review B. 1997;**55**:16293

[23] Bernardes E et al. Spin-orbit interaction in symmetric wells with two subbands. Physical Review Letters. 2007;**90**:076603

[24] Calsaverini RS et al. Intersubband-induced spin-orbit interaction in quantum wells. Physical Review B. 2008;**78**:155313

[25] Fu JY et al. Persistent skyrmion lattice of noninteracting electrons with spin-orbit coupling. Physical Review Letters. 2016;**117**:226401

[26] Fu JY, Carlos Egues J. Spin-orbit interaction in GaAs wells: From one to two subbands. Physical Review B. 2015; **91**:075408

[27] Goni AR, Haboeck U, Thomsen C, Eberl K, Reboredo FA, Proetto CR, et al. Exchange instability of the two-dimensional electron gas in semiconductor quantum wells. Physical Review B. 2002;**65**:121313(R)

[28] Rigamonti S, Proetto CR. Signatures of discontinuity in the exchange-correlation energy functional derived from the subband electronic structure of semiconductor quantum wells. Physical Review Letters. 2007;**98**:066806

[29] Vurgaftman I et al. Band parameters for III-V compound semiconductors and their alloys. Journal of Applied Physics. 2001;**89**:5815

[30] Yi W et al. Bandgap and band offsets determination of semiconductor heterostructures using three-terminal ballistic carrier spectroscopy. Applied Physics Letters. 2009;**95**:112102

[31] Walser MP et al. Dependence of the Dresselhaus spin-orbit interaction on the quantum well width. Physical Review B. 2012;**86**:195309

[32] Wang W et al. Two distinct regimes for the electrical control of the spin–orbit interaction in GaAs wells. Journal of Magnetism and Magnetic Materials. 2016;**411**:84

[33] Kunihashi Y et al. Proposal of spin complementary field effect transistor. Applied Physics Letters. 2012;**100**: 113502

[34] Sheng XL et al. Topological insulator to Dirac semimetal transition driven by sign change of spin-orbit coupling in thallium nitride. Physical Review B. 2014;**90**:245308

[35] Fletcher R et al. Two-band electron transport in a double quantum well. Physical Review B. 2005;**71**:155310

[36] Glazov MM et al. Two-dimensional electron gas with spin-orbit coupling disorde. Physica E. 2010;**42**:2157

[37] Morgenstern M et al. Scanning tunneling microscopy of two-dimensional semiconductors: Spin properties and disorder. Physica E. 2012; **44**:1795

[38] Glazov MM, Sherman EY. Nonexponential spin relaxation in magnetic fields in quantum wells with random spin-orbit coupling. Physical Review B. 2005;**71**:241312(R)

[39] Sherman EY. Random spin-orbit coupling and spin relaxation in symmetric quantum wells. Applied Physics Letters. 2003;**82**:209

[40] Wang W, Fu JY. Electrical control of the spin-orbit coupling in GaAs from single to double and triple wells. Superlattices and Microstructures. 2015; **88**:43

[41] D'yakonov MI, Perel' VI. Spin orientation of electrons associated with the interband absorption of light in semiconductors. Soviet Physics–JETP. 1971;**33**:1053

[42] Meier F, Zakharchenya BP. Optical Orientation. Amsterdam: North-Holland; 1984

Chapter 6

Evolution of Nanometer-to-Micrometer Grain Size in Multiferroic Properties of Polycrystalline Holmium and Yttrium Manganite

Nor Hapishah Abdullah, Raba'ah Syahidah Azis, Muhammad Syazwan Mustaffa, Mohd Nizar Hamidon and Farah Nabilah Shafiee

Abstract

The parallel evolution of microstructure development via grain size changes from a nano-to-micron size regime toward multiferroic property development has been established in this research work. This kind of observation is not present in the literature in this research area, and studies of the link between morphological properties and ferroelectric properties of multiferroic materials have been focusing solely on the product of the ultimate sintering temperature, mostly neglecting the parallel evolutions of morphological properties and their relationship at varied chosen sintering temperatures. Holmium manganese oxide and yttrium manganese oxide were both prepared via high-energy ball milling (HEBM) in a hardened steel vial for 12 h. The pressed pellet went through multi-sample sintering, whereas the samples were sintered starting from 600 to 1250°C with 50°C increments for any one sample being subjected to only one sintering temperature. Orthorhombic $HoMn_2O_5$ and YMn_2O_5 phases were observed to exist in both as milled powder. The degree of crystallinity increased with increasing sintering temperature. Hexagonal $HoMnO_3$ peaks were observed for sintering temperature ≥1050°C. As for $YMnO_3$ series, the single phase of hexagonal $YMnO_3$ started to appear at sintering temperature ≥1000°C. FESEM micrographs revealed that as the sintering temperature increased, the grain size increased, consequently increasing the geometric ferroelectric behavior. The polarization-electric field (*P-E*) plot reveals that $HoMnO_3$ and $YMnO_3$ are highly leaky ferroelectrics with a *P-E* curve shape different from the normal shape of highly insulating ferroelectrics. It shows that the remanent polarization and electric field increased generally with increasing grain size. For both series, there existed a difference based on their difference of crystallinity, microstructure data, and phase purity changes. Larger grain size is known to give ease for polarization to take place.

Keywords: multiferroic, microstructure, holmium manganite (HMO), yttrium manganite (YMO), sintering temperature

IntechOpen

1. Introduction of hexagonal manganites as multiferroic materials

Multiferroic materials consist of more than one ferroic polarization, and the term multiferroic was first coined by Schmid in 1994 to indicate a material that has either two or three different kinds of ferroic orders like ferromagnetism, ferroelectricity, ferroelasticity, and ferrotoroidicity in the same phase [1]. The definition is often extended to antiferroic orderings. Ferroic materials are defined as possessing a spontaneous order, the direction of which can be switched by using an external field. A ferromagnet has a spontaneous magnetization M and shows hysteresis under an applied magnetic field H. A ferroelectric has a spontaneous polarization P and shows hysteresis under an applied electric field E. As for ferroelastic, it has a spontaneous strain and shows hysteresis under an applied stress. Such spontaneous order typically occurs as a result of a phase transition when the material is cooled below a particular temperature. Following Schmid in 1994, these kinds of materials are recognized as multiferroics (**Figure 1**). Nowadays, what most people mean by multiferroic predominantly applies to the coexistence of magnetism and ferroelectricity.

In the past 5 years, research into multiferroics has branched into several different areas. The use of multiferroics for technology is currently a widely researched field, and the fundamental physics governing the strong magnetoelectric coupling seen in multiferroics is still not fully understood. Now that the causes of multiferroic behavior are better understood, it has been possible for us to observe the evolution of their microstructure with respect to their ferroelectric behavior acting in the nanoscale-to-micron-sized grain regimes. As for the chosen materials, the hexagonal rare-earth manganites, $RMnO_3$, were first discovered in 1963 [2–4]. Hexagonal $RMnO_3$ compounds show a strong ferroelectric ordering with saturated polarization larger than 5.6 μCcm^{-2} [5–7]. Thus, hexagonal rare-earth manganites are classified as an interesting family with multiferroic properties, which is also the driving force for them to be the focus of this study.

Bertaut et al. in 1963 discovered the ferroelectricity in hexagonal $REMnO_3$ [3, 4]. Precise structural investigations have been employed to find the origin of ferroelectricity in hexagonal rare-earth manganites since its discovery by Bertaut et al. [8–11]. Consequently, the origin of ferroelectricity is geometrically driven by the displacement between RE^{3+} and O^{2-} resulting from a structural phase transition. A structural phase transition from a high-temperature paraelectric phase (space group $P6_3/mmc$) to a low-temperature ferroelectric phase (space group $P6_3cm$) has been observed in hexagonal $REMnO_3$. $YMnO_3$ may serve as a representative of all hexagonal rare-earth manganite systems, especially for $HoMnO_3$ with its similar size of the rare-earth ion and lattice parameters.

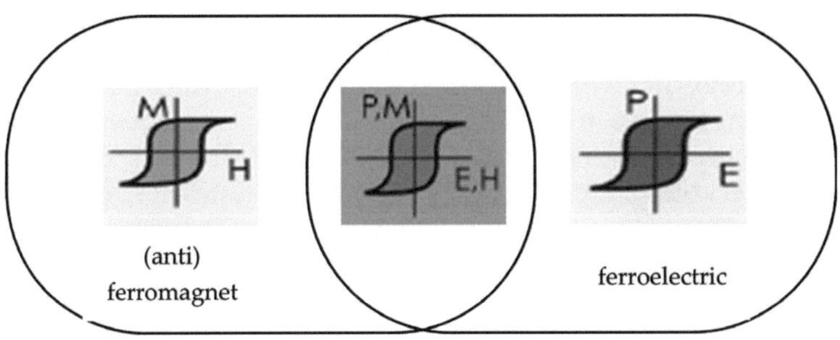

Figure 1.
Multiferroics combine the properties of ferroelectrics and magnets.

2. Brief overview of the synthesis method

Polycrystalline holmium and yttrium manganite samples were synthesized by a solid state reaction via high-energy ball milling (HEBM) with 12 h milling time. This was carried out by mixing and milling together according to the stoichiometric ratios of the required metal oxide powders, followed by pelletizing and furnace heating. The most important measurements involved in this project are ferroelectric parameters with the evolution of their microstructure. These are, ferroelectric hysteresis *P-E* loop along with microstructure analysis by FESEM measurement. For the purpose of the multiferroic material preparation in this project, the following starting materials were used:

- Holmium oxide (Ho_2O_3), 99.99%, Strem Chemicals
- Yttrium oxide (Y_2O_3), 99.99%, Alfa Aesar
- Manganese (III) oxide (Mn_2O_3), 99%, Strem Chemicals

The chemical equation via mechanical alloying (high-energy ball milling) for both desired samples are shown below:

- $0.5\ Ho_2O_3 + 0.5\ Mn_2O_3 \rightarrow HoMnO_3$
- $0.5\ Y_2O_3 + 0.5\ Mn_2O_3 \rightarrow YMnO_3$

The requirement for the sample preparation is to have the powder materials in the form of nanosized starting particles. With the use of HEBM, the duration of milling time has been chosen to be 12 h, which is optimum in order to obtain nanosized particles. The mixed material was crushed by using high-energy ball milling (HEBM) in order to facilitate the solid state reaction. The mechanical alloying (MA) process starts with mixing the powders in the right proportion and loading them into the mill along with the grinding medium (usually steel balls). The mixture is then milled for the desired length of time until a steady state is reached when the composition of every powder sample contains the same proportion of the elements in the starting powder mix. The milled powder is then transformed into a bulk shape and heat-treated to obtain the desired microstructure and properties. The key of the process involved is the raw materials, the mill, and the process variables [12]. The selected optimum parameters for the milling and ball-to-powder ratio (BPR) were 12 h and 10:1, respectively, and were used for the preparation of all the multiferroic nanoparticles used in this project. For pellet preparation, ~1.0 g is required from the as-milled powder for each sintering temperature. The mechanically alloyed nanoparticle materials were weighed according to the calculated formula using an analytical microbalance (A&D, model GR-200) and granulated by using 2% PVA. The samples were then lubricated with zinc stearate in order to reduce the density gradient caused by friction of the powder along the wall of the mold. The transformation of the previously granulated powder into a pellet shape was carried out by pressing the mold with a force of 200 MPa by using a pressing machine. Suitable pressure is important in order to obtain a uniform density compact resulting in more uniform shrinkage during sintering.

Sintering is a heating process in which the atomic mobility of the compact is sufficient to permit the decrease of the free energy associated with the grain boundaries [13]. It is the most critical and expensive process step as it yields the required crystal structure, oxidation state, microstructure, and physical condition of a material. The purpose of the sintering process is to complete the interdiffusion of the component metal ions into the desired crystal lattices and to develop the polycrystalline microstructure at as low a temperature as possible, first to ensure a high grade of microstructural homogeneity, second to avoid cannibal grain growth, and third to save energy cost. At this stage, atoms are closely in contact with each other, hence reducing the space between them. Therefore, shrinkage occurs as a consequence of aggregation of the entire substance. As required by commercial production practice, the desired shrinkage is between 10 and 20%.

Since the aim of this research is to study in detail the evolution of the microstructure and the subsequent effect of the ferroelectric processes, sintering was used as an agent to control such desired changes that would be observed in the microstructure. The sintering temperature used for the purpose of detecting the sintering effects on the *P-E* hysteresis loops was from 600 to 1250°C with the increase of 50°C giving a total of 14 samples for each batch. They were sintered separately for 10 h in an electric furnace. The average dimensions of the final product with the pellet shape were about 10 mm diameter and 1–2 mm thickness.

3. Grain size evolution and multiferroic properties of holmium and yttrium manganite

Figure 2(**a**) and (**b**) shows the micrograph of the holmium and yttrium manganite powder samples milled for 12 h, which have been measured by using transmission electron microscopy (TEM) for particle size confirmation. The particle size was measured by taking diameters of 200 particles, and it was found to have inhomogeneity variation from 7 to 70 nm with an average around 27 nm (a) and 11–87 nm with average particle sizes around 30 nm (b) for holmium and yttrium manganite, respectively. The wide range of particle size distribution was due to the nonuniformity of the force between the balls and the vial during the milling process. In order to obtain a uniform grain size, particle size is an important quantity that needs to be considered. The agglomerations of the particles also occurred due to the large surface area and being subjected to repeated cold welding and fracturing during the HEBM process. This particle size is an average value considering that the particles are spherical. The particles are known to have higher reactivity due to the larger surface area. The surface atoms are more unstable (and reactive). This instability is related to their position on the lattice that forces them to unbind to their neighboring atoms or molecules. For the case of nanoparticles, as the ratio of surface/bulk atoms increases, the instability (and reactivity) also increases. Higher

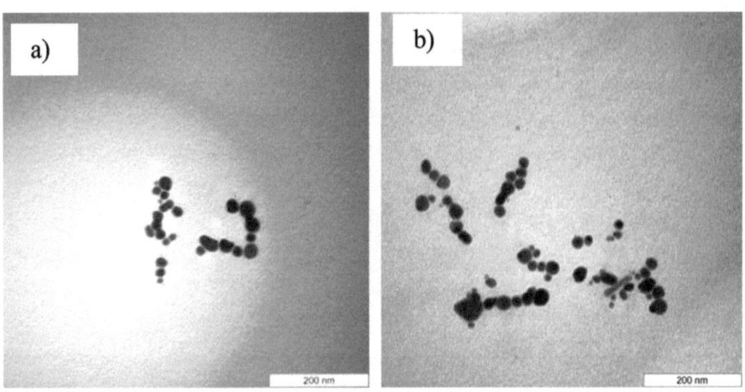

Figure 2.
TEM micrograph of resulting particles after 12 h of milling time for (a) HMO (b) YMO.

reactivity in the starting powder is required in order to prepare a series of samples in which the observation of nanosized starting particles followed by nano-to-micron size grains is required in this study.

A control of the crystal orientation is significantly important, because the ferroelectric polarization of $HoMnO_3$ and $YMnO_3$ appears along the hexagonal *c*-axis. **Figure 3(a)** and **(b)** reveals the XRD spectra of HMO and YMO samples, respectively. At 12 h high-energy ball milling, broad diffraction peaks belonging to

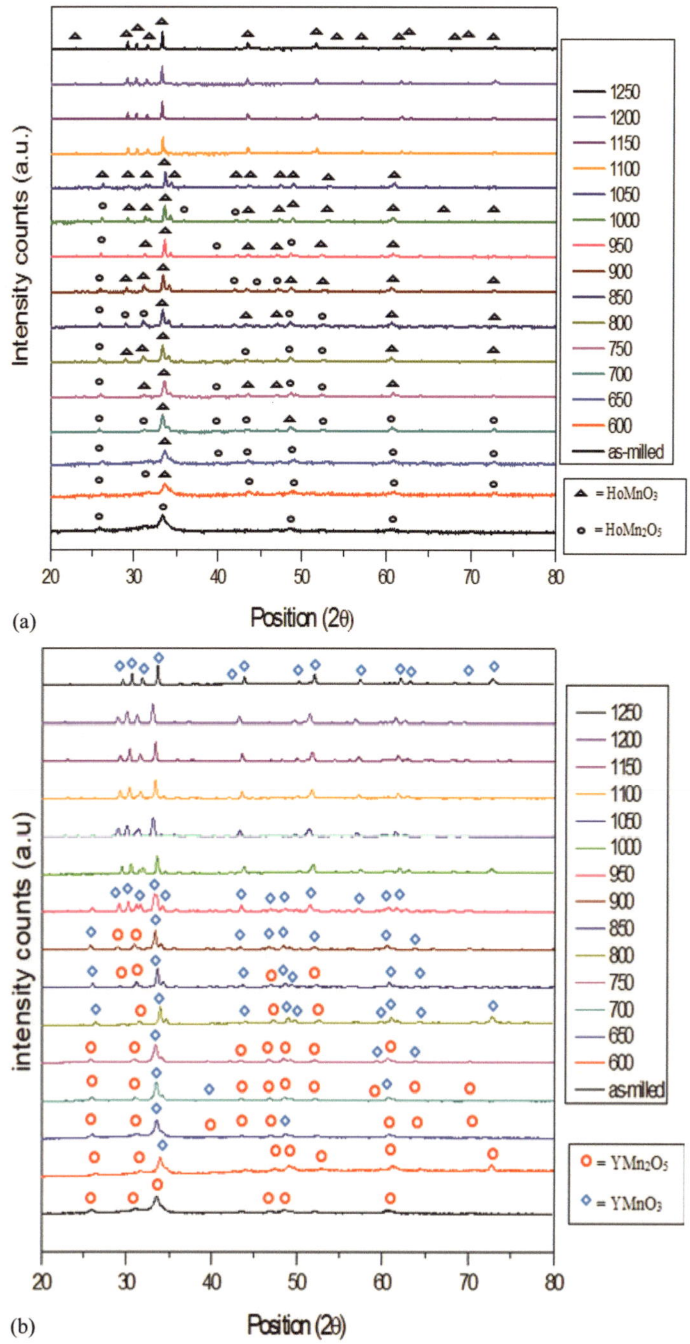

Figure 3.
(a) XRD patterns of for as milled continued with sintering from 600 to 1250°C for HMO. (b) XRD patterns of for as milled continued with sintering from 600 to 1250°C for YMO.

orthorhombic $HoMn_2O_5$ and YMn_2O_5 were observed to exist, which means that fully amorphous powders were formed. Hexagonal $HoMnO_3$ and $YMnO_3$ peaks were not observed in the as-milled powder due to insufficient thermal energy to form the structure. Impurity phases other than raw powders were not detected in the as-prepared sample. However, high-energy ball milling facilitates the reaction between the raw materials where the sintering temperature could be reduced to lower than that normally required in the conventional method due to higher surface reactivity [14]. The energy is transferred from the milling media to the powder particles by the continuous fracture and cold welding process. As the grain size started to increase, small peaks of hexagonal $HoMnO_3$ and $YMnO_3$ were detected even with fairly low intensity. Starting from 600°C sintering temperature up to 1000°C, both peaks consisting of orthorhombic $HoMn_2O_5$/ YMn_2O_5 and hexagonal $HoMnO_3$/$YMnO_3$ intensity were observed to decrease and increase, respectively. As the sintering temperature continued to increase, the secondary phase reduced in intensity. As can be seen in the figure, the main peak in each sintering temperature from 600 to 1000°C exists a splitting which is the evidence of the formation of a new structure from orthorhombic to hexagonal. The transformation from orthorhombic to a single-phase hexagonal structure started to occur at 1050°C for the HMO sample and at 1000°C for the YMO sample. Further increasing the sintering temperature from 1050 to 1250°C raised the intensity of the major peaks belonging to hexagonal holmium and yttrium manganese oxide ($HoMnO_3$ and $YMnO_3$), which demonstrated the improvement in the degree of crystallinity of the sintered samples besides releasing the strain induced by milling. The results demonstrate that the hexagonal phase of $HoMnO_3$ and $YMnO_3$ could not be formed only by milling since the energy imparted by the collision of the milling media to the starting powders is not sufficient to increase the reactivity between the particles. Besides, without the sintering process, insufficient energy will be provided to stimulate the reaction between particles. As the sintering temperature increases, a larger grain size would be formed due to the combination of grains, thus inducing the formation of a more crystalline phase with a hexagonal structure. Furthermore, a ferroelectric phase would be formed because of the increasing electric dipole moment of the crystalline hexagonal structure, whereas it is known that ferroelectricity is geometrically driven by the displacement between RE^{3+} and O^{2-} as a result of a structural phase transition [8–11]. Therefore, a higher sintering temperature, say 1250°C, would lead to a higher value of dielectric constant and ferroelectric order. At lower sintering temperature, the sample exhibited a smaller grain size that will contribute a larger amount of amorphous grains compared to samples with a larger grain size. This amount was significant in view of the fact that the grain boundary volume was not negligible for the nanometer grain size [15].

The microstructural images of the sintered pellets were obtained using a Nova NanoSEM 50 scanning electron microscope. The SEM micrographs of the HMO and YMO samples are shown in **Figures 4(a)–(n)** and **5(a)–(n)**. In the HMO samples, the sintering temperature from 600 to 900°C with a grain size of 30 to 65 nm determines the slow rate of grain growth, while at 950, 1000, and 1050°C (from 108 to 296 nm), it indicates moderate increment. The grain growth was observed to be increased at 1100 up to 1250°C (854 nm–2.9 μm) due to creation of a pure single hexagonal HMO phase. As for the YMO samples, grain growth occurred at 600–900°C (from 49 to 75 nm) and continued by adequate increment at 950, 1000, and 1050°C (from 82 to 314 nm). The grain growth was observed to be increased at 1100°C up to 1250°C (543 nm–2.1 μm). The results demonstrate that multi-sample sintering involved a transition from the slow-moderate-rapid grain growth process. The grains grew at the expense of others and were created from movement of grain boundaries and grains, respectively. The pores will form an interconnected channel along grain edges when

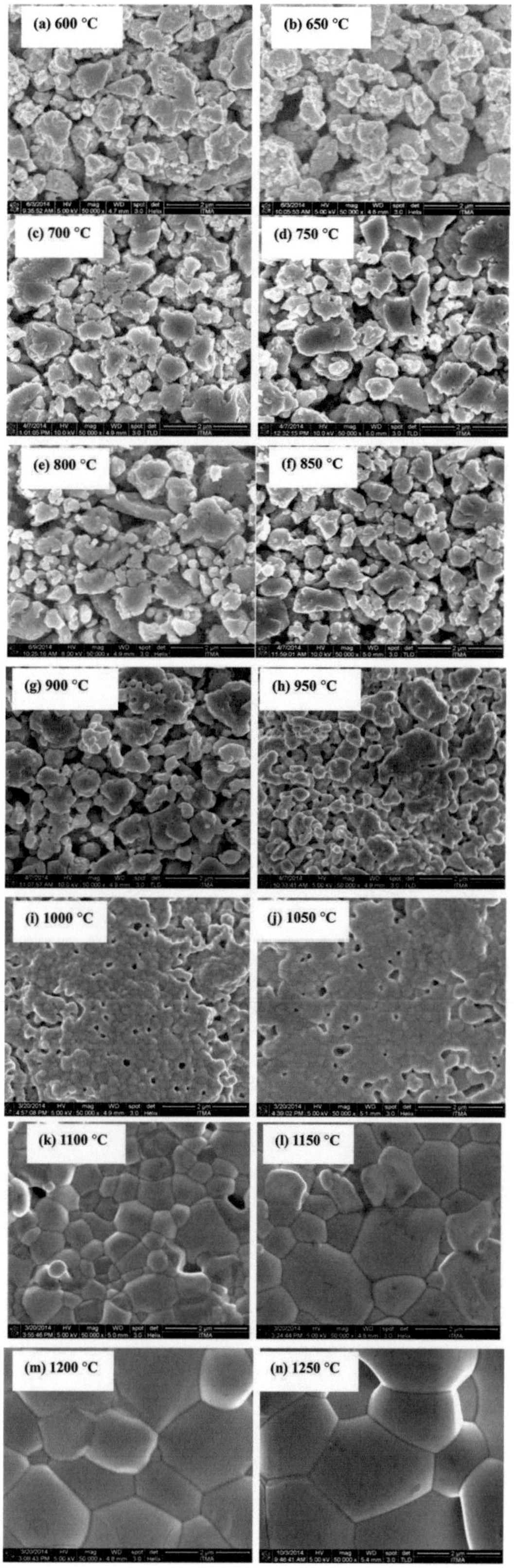

Figure 4.
(a–n) FESEM micrograph from 600 to 1250°C for HMO.

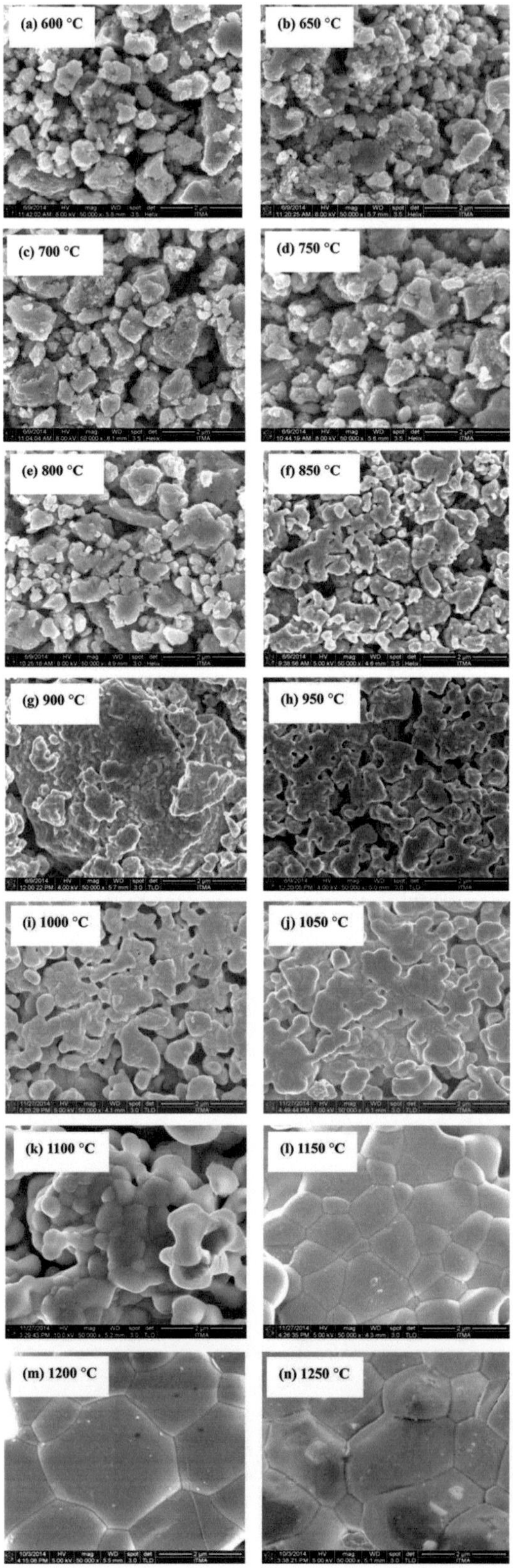

Figure 5.
(a–n) FESEM micrograph of YMO sintered from 600 to 1250°C.

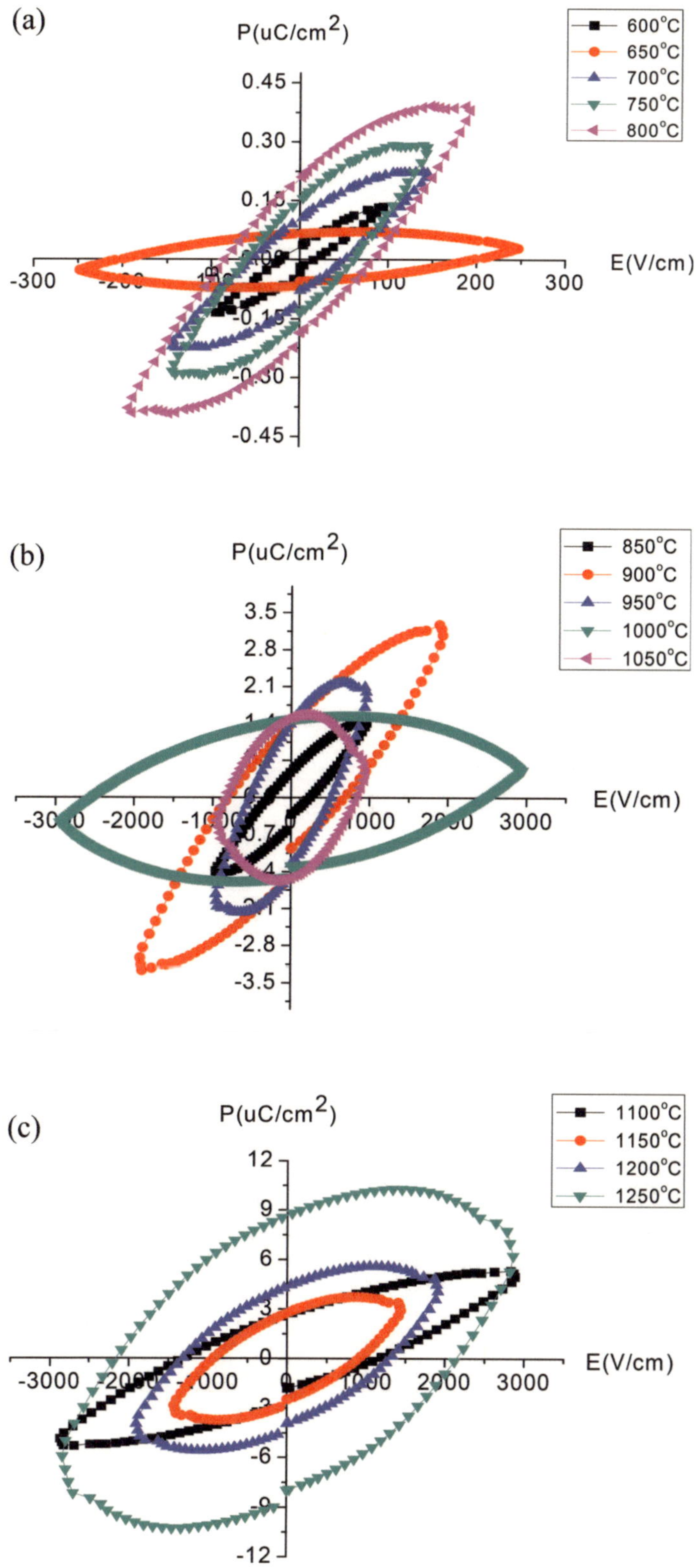

Figure 6.
(a) P-E for HMO samples sintered at 600–800°C. (b) P-E for HMO samples sintered at 850–1050°C. (c) P-E for HMO samples sintered at 1100–1250°C.

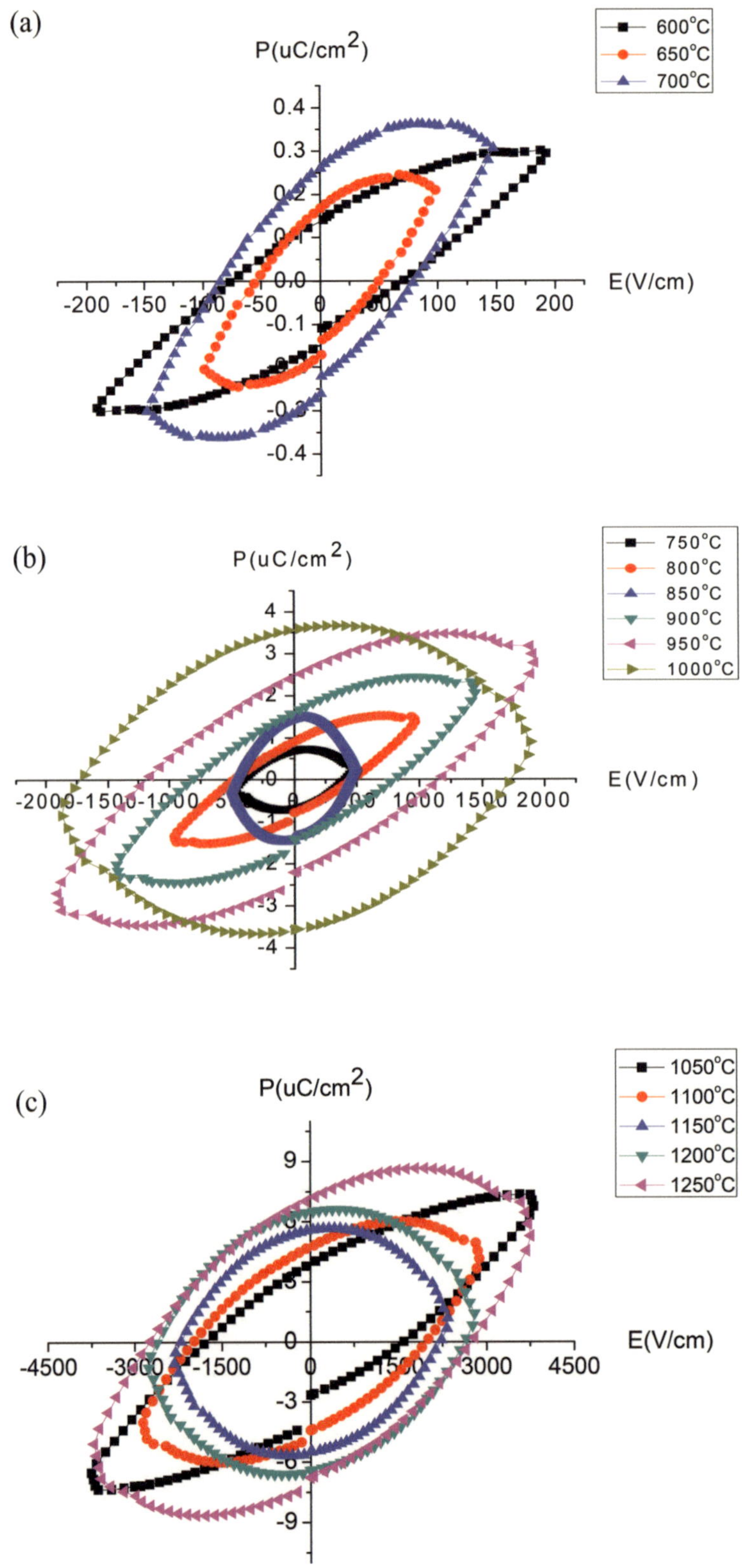

Figure 7.
(a) P-E for YMO samples sintered at 600–700°C. (b) P-E for YMO samples sintered at 750–1000°C. (c) P-E for YMO samples sintered at 1050–1250°C

there is formation of a necking process between particles in the powder compacts. However, the pore channels were disconnected and isolated when the sintering process was introduced. Microstructural interpretations at higher sintering temperatures from 1100 to 1250°C in **Figures 4(k)–(n)** and **5(k)–(n)** correspond to the final stage of sintering. The diffusion process of vacancies from the pores along grain boundaries will cause the pores grew to be closed and have been slowly eradicated with a slight densification. The grains with a hexagonal structure could be perceived in this stage as the sintering temperature and grain size increased.

Figures 6(a)–(c) and **7(a)–(c)** show the polarization induced by applying an electric field for all sintered samples of HMO and YMO, respectively. Lower sintering temperatures, 600–800°C for the HMO *PE* loop (**Figure 6a**) and 600–700°C for YMO (**Figure 7a**), reveal a lossy capacitor response with low P_r, indicating the major paraelectric behavior contributed by a large amount of orthorhombic phase with very little contribution of the hexagonal phase. It is speculated that the presence of the second phase (shown in the XRD spectra) makes the electron to be detached and free as free charges which contributed to cause the loop to leak. As the sintering temperature increased from 850 to 1050°C for HMO (**Figure 6b**) and 750 to 1000° C for YMO (**Figure 7b**), an ideal resistor response curve with higher values of P_r and E_c was obtained. This is due to the existence of the geometric ferroelectric derived from the hexagonal phase, which started to dominate as the grain size increased with the increase of sintering temperature. The grain boundary volume in a smaller grain size is higher compared to a larger grain size, which was the reason why the P_r got better as the grain size increased. The effect of grain boundary on polarization includes two factors. The first one is, the grain boundary is a low-permittivity region, which means that the grain boundary has poor ferroelectricity with little or even no polarization. Secondly, space charges in the grain boundary exclude polarization charge on the grain surface, which would form a depletion layer on the grain surface, which results in polarization discontinuity on the grain surface to form a depolarization field and the polarization decreases. At higher sintering temperature as seen in **Figure 6c** for HMO and **Figure 7c** for YMO, it was confirmed that the contribution of the majority hexagonal phase has improved the geometric ferroelectric behavior. Sintering at higher temperature shows the increase of the average grain size, which gave a good response in the polarization hysteresis loop. The contribution of larger grains reduced the secondary peaks while having high crystallinity and enhancing the response of the electric field. As shown in the polarization hysteresis (**Figures 6** and 7), applying the electric field increases the polarization values but still showed that the *P-E* hysteresis loops are not saturated. HMO and YMO are known to behave as a leaky ferroelectric as mentioned in previous reports [16, 17]. As can be seen in **Figures 6** and 7, the shapes of the hysteresis loop are mainly related to the combination of an ideal resistor and lossy capacitor response. The mobility of free charges will also contribute to higher conductivity and cause higher leaking current, which would be the main reasons for the E_c values to be unpredictable. Even though the hysteresis loops are seen to be leaky in nature, they are still significant enough to definitely confirm the ferroelectricity of the HMO and YMO samples.

4. Conclusion

$HoMnO_3$ and $YMnO_3$ multiferroics were successfully synthesized via high-energy ball milling, and the parallel evolution of ferroelectric with microstructural properties has been studied. High-energy ball milling was able to produce high-degree crystallinity and some trends of electrical properties due to the high surface reactivity of starting materials. It must be noted that this clear understanding has been made possible only through a progressive sintering scheme of nanometer-particle compacts from an unusually low sintering temperature (600°C) to a somewhat high sintering temperature (1250°C). From the discussion presented, the evolution of a nanometer-to-micron grain size regime has been presented and the changing patterns of ferroelectrics are now clearly understood. It was found that the ferroelectric and magnetic properties generally correlated with intrinsic and extrinsic properties. The intrinsic contribution came from the contribution of the crystal structure in which orthorhombic, hexagonal, and more or less orthorhombic + hexagonal phases affected the phase that will contribute to the proper behavior of the ferroelectric. The effects of grain size of the two series of manganites have been observed by the scheme of nanosized starting particles followed by nano-to-micron grain sized regime data. Microstructural changes revealed a revolution of the crystal structure from orthorhombic to hexagonal at a larger grain size regime. The ferroelectric behavior was also observed to change with the change of microstructure along with the structural transformation from orthorhombic to hexagonal. For a general conclusion, the intrinsic effect occurred in the low sintering temperature region (600–1000°C for $HoMnO_3$) and (600–900°C for $YMnO_3$). The property changes at this region are due to crystal structure transformation. The extrinsic effect was more obvious at higher sintering temperature that is 1050–1250°C for $HoMnO_3$ and 950–1250°C for $YMnO_3$ in the hexagonal structure. The optimum condition to obtain a sample with very fine properties could be obtained by performing high-energy ball milling for 12 h followed by sintering at 1250°C with 10 h holding time. These steps were required in order to reach a very stable hexagonal structure for most advantageous ferroelectric and magnetic properties. The study of the evolution work has resulted in greater appreciation of the theoretical and experimental difficulties involved, if not in new knowledge of the behavior of multiferroic studies in evolution. In fact, there were no reported studies regarding these evolution works in the multiferroic field.

Acknowledgements

The authors are thankful to the Materials Synthesis and Characterization Laboratory (MSCL), Functional Devices Laboratory (FDL), Institute of Advanced Technology (ITMA), and also the Department of Physics, Faculty of Science, Universiti Putra Malaysia (UPM), for the measurement facilities.

Author details

Nor Hapishah Abdullah[1], Raba'ah Syahidah Azis[2,3]*,
Muhammad Syazwan Mustaffa[2], Mohd Nizar Hamidon[1] and Farah Nabilah Shafiee[3]

1 Functional Devices Laboratory (FDL), Institute of Advanced Technology, Universiti Putra Malaysia, UPM, Serdang, Selangor, Malaysia

2 Department of Physics, Faculty of Science, Universiti Putra Malaysia, UPM, Serdang, Selangor, Malaysia

3 Materials Synthesis and Characterization Laboratory (MSCL), Institute of Advanced Technology, Universiti Putra Malaysia, UPM, Serdang, Selangor, Malaysia

*Address all correspondence to: rabaah@upm.edu.my

References

[1] Schmid H. Multiferroic magnetoelectrics. Ferroelectrics. 1994; **162**:317-338

[2] Subba Rao GV, Chandrashekhar GV, Rao CNR. Are rare earth orthochromites ferroelectric?. Solid State Communications. 1968;**6**(3):177-179

[3] Kovalev AV, Andreeva GT. Low temperature X-ray study of $Ni_3B_7O_{13}I$ boracite. Comptes Rendus Academy of Science. 1956;**2**:56

[4] Coeure P, Guinet F, Peuzin JC, Buisson G, Bertaut EF. Ferroelectric properties of hexagonal orthomanganites of yttrium and rare earths. Proceeding International Meeting Ferroelectric (Prague). 1966;**1**: 332

[5] Katsufuji T, Mori S, Masaki M, Moritomo Y, Yamamoto N, Takagi H. Dielectric and magnetic anomalies and spin frustration in hexagonal $RMnO_3$ (R=Y, Yb, and Lu). Physical Review B. 2001;**64**:1-6

[6] Lottermoser T, Lonkai T, Amann U, Hohlwein D, Ihringer J, Fiebig M. Magnetic phase control by an electric field. Nature. 2004;**430**:541-544

[7] Fiebig M, Lottermoser T, Fröhlich D, Goltsev AV, Pisarev RV. Observation of coupled magnetic and electric domains. Nature. 2002;**419**:818-820

[8] Yakel HL, Koehler WD, Bertaut EF, Forrat F. On the crystal structure of the manganese (III) trioxides of the heavy lanthanides and yttrium. Acta Cryst. 1963;**16**:957

[9] Aken BBV, Bos JWG, Groot RA, Palstra TTM. Asymmetry of electron and hole doping in $YMnO_3$. Physical Review B. 2001;**63**:125-127

[10] Aken BBV, Palstra TTM, Filippetti A, Spaldin NA. The origin of ferroelectricity in magnetoelectric $YMnO_3$. Nature Materials. 2004;**3**: 164-170

[11] N'enert G. Orbital ordering and multiferroics [PhD thesis]. The Netherlands: University of Groningen. Vol. 55–59; 2007. pp. 93-96

[12] Suryanarayana C. Mechanical alloying and milling. Progress in Materials Science. 2001;**46**:1-184

[13] Slick PI. Ferromagnetic Materials. Vol. 2. The Netherlands: North-Holland Publishing Company; 1980. p. 209

[14] Waje SB, Hashim M, WanYusoff WD, Abbas Z. Sintering temperature dependence of room temperature magnetic and dielectric properties of $Co_{0.5}Zn_{0.5}Fe_2O_4$ prepared using mechanically alloyed nanoparticles. Journal of Magnetism and Magnetic Materials. 2010;**322**:686-691

[15] Idza IR, Hashim M, Rodziah N, Ismayadi I, Norailiana AR. Influence of evolving microstructure on magnetic-hysteresis characteristics in polycrystalline nickel-zinc ferrite, $Ni_{0.3}Zn_{0.7}Fe_2O_4$. Materials Research Bulletin. 2012;**47**:1345-1352

[16] Sahu JR, Ghosh A, Rao CNR. Multiferroic properties of $ErMnO_3$. Materials Research Bulletin. 2009;**44**: 2123-2126

[17] Scott JF. Ferroelectric go bananas. Journal of Physics: Condensed Matter. 2008;**20**(2):1-2